Cyclodextrins

Preparation and Application in Industry

Cyclodextrins

Preparation and Application in Industry

Edited by

Zhengyu Jin
Jiangnan University, China

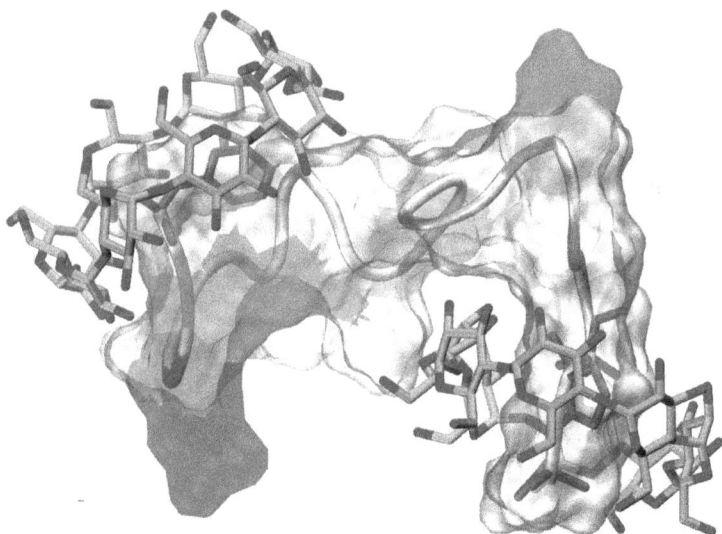

World Scientific

NEW JERSEY · LONDON · SINGAPORE · BEIJING · SHANGHAI · HONG KONG · TAIPEI · CHENNAI · TOKYO

Published by

World Scientific Publishing Co. Pte. Ltd.

5 Toh Tuck Link, Singapore 596224

USA office: 27 Warren Street, Suite 401-402, Hackensack, NJ 07601

UK office: 57 Shelton Street, Covent Garden, London WC2H 9HE

Library of Congress Cataloging-in-Publication Data

Names: Jin, Zhengyu, 1960–   author.

Title: Cyclodextrins : preparation and application in industry / Zhengyu Jin
    (Jiangnan University, China).

Description: New Jersey : World Scientific, 2017. | Includes bibliographical references and index.

Identifiers: LCCN 2017037693 | ISBN 9789813229655 (hardcover : alk. paper)

Subjects: LCSH: Cyclodextrins. | Cyclodextrins--Industrial applications. |
    Oligosaccharides. | Supramolecular chemistry.

Classification: LCC TP248.C94 J56 2017 | DDC 572/.565--dc23

LC record available at https://lccn.loc.gov/2017037693

British Library Cataloguing-in-Publication Data

A catalogue record for this book is available from the British Library.

For any available supplementary material, please visit
http://www.worldscientific.com/worldscibooks/10.1142/10701#t=suppl

Typeset by Stallion Press
Email: enquiries@stallionpress.com

Contents

1. Introduction

Junrong Huang, Qi Yang, and Huayin Pu

*Shaanxi University of Science and Technology,
Weiyang District, Xi'an, Shaanxi 710021, China*

1.1 Introduction

Cyclodextrins (CDs) are cyclic, nonreducing oligosaccharides linked through α-1,4 glycosidic bonds. Due to steric repulsion, CDs have more than six glucose units. α-, β-, and γ-CDs contain six, seven, and eight glucose units, respectively. The chair conformation of glucopyranose units results in CD molecules having a truncated cone shape, with a somewhat hydrophobic central cavity and a hydrophilic outer surface [1]. This structure enables them to interact with poorly water-soluble compounds, thereby solubilizing them via the formation of host–guest inclusion complexes [2]. Normally, the hydrophobic chains of amphiphiles are included in host–guest complexes with 1:1 or 2:1 stoichiometry with high binding constants. In the late 1960s, the appearance of polyether and realization of its recognition capacity created the concept of host–guest chemistry. CDs possess excellent properties in terms of molecular recognition, molecular interaction, and molecular aggregation, and many molecules of a suitable size can undergo inclusion complexation with CDs. The wide availability and low cost of CDs facilitates their use in various fields including analysis, catalysis, and surface chemistry, and in many industries such as pharmaceuticals, cosmetics, textiles, and food [3].

In recent times, supramolecular chemistry has attracted considerable attention as an important subdiscipline of chemistry that develops molecular building blocks for the construction of novel systems with intriguing properties that differ from their separate components [4]. CDs are generally composed of both hydrophilic and hydrophobic moieties. Through hydrophobic and other noncovalent interactions, CD molecules can form various self-assembled structures in aqueous solutions including micelles, vesicles, lyotropic liquid crystals, and gels, all of which have found many applications in the fields of cosmetics, drug delivery, materials synthesis, and microreactors. A great deal of attention has been given to the construction of novel ordered and functional assemblies due to their delicate and highly organized aggregate structures. Different approaches have been developed including molecular structure modification to tune the amphiphile hydrophilic/hydrophobic balance and introduce molecular motifs such as host molecules to form inclusion complexes, and CDs are now considered effective modulators of the self-assembly of amphiphiles [5].

Host–guest interactions of CDs are derived from many aspects including hydrophobic interactions, Van der Waals interactions, the ring tension of the CD, the surface tension of the solvent, and the effect of hydrogen bonds. Numerous guest molecules of an appropriate size are generally able to form inclusion complexes with CDs (Fig. 1.1) [6, 7].

Polymeric systems based on CDs and certain guest molecules have been developed in recent years. For example, the application of nanoparticle and micelle macromolecular materials has attracted widespread attention in medical and biological fields, especially for continuous drug release, targeted delivery, and tissue engineering. Many stimuli-responsive polymeric networks with novel structures

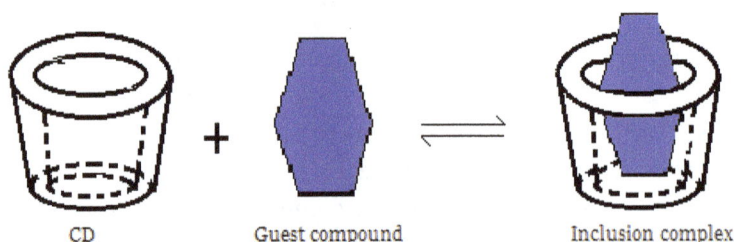

CD   Guest compound   Inclusion complex

Fig. 1.1.   Schematic representation of inclusion complex formation (1:1) between a CD and a guest molecule [6].

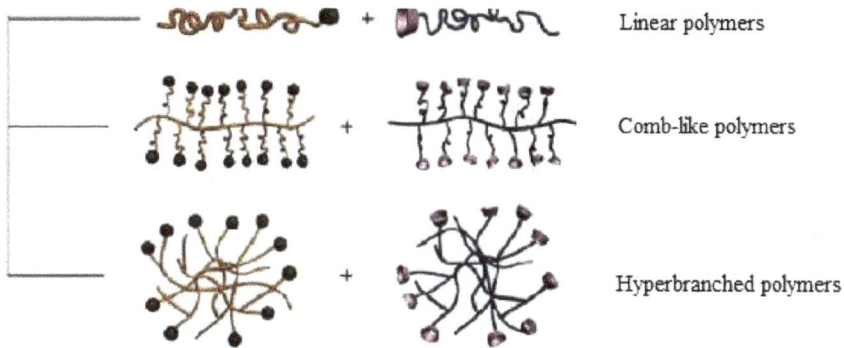

Fig. 1.2. Schematic illustration of polymeric systems based on CDs and guest molecule inclusion complexes [3].

have been designed through chemical cross-linking, physical aggregation, and other means [3]. Polymeric systems based on CDs and guest molecule inclusion complexes can vary in structure and may be linear, branched, comb-like, or hyperbranched, as shown in Fig. 1.2. These structures can subsequently form higher-order structures such as peels, micelles, vesicles, and tubes, which can be used in many ways.

Supramolecular systems often resemble naturally occurring molecules in structure and function; hence they are considered biomimetics. Key life processes including photosynthesis and oxygen storage are dependent on complex formation via supramolecular building molecules such as porphyrinoids (Fig. 1.3) [4].

1.2 Applications in foods

1.2.1 *Applications in controlled flavor release*

Flavor plays an important role in consumer satisfaction and influences the consumption of foods. Manufacturing and storage processes, packaging materials and ingredients in foods often modify the overall flavor by reducing aroma compound intensity or generating off-flavor components [8]. To limit aroma degradation or loss during processing and storage, it is beneficial to encapsulate volatile ingredients prior to use in foods, especially in beverages [8, 9]. The addition of β-CD as a stabilizing or thickening agent can help to retain some aroma

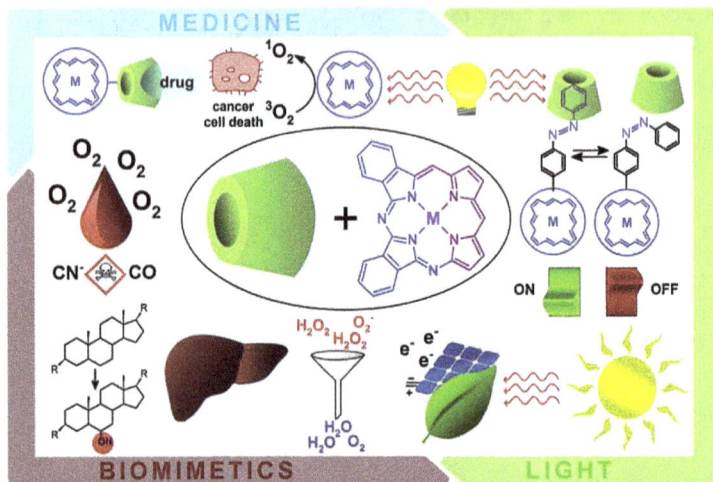

Fig. 1.3.   Various functionalities of cyclodextrin-porphyrinoid systems [4].

compounds in food matrices during cooking, pasteurization, and other thermal processes [10].

1.2.2 *Applications in special-flavor correction*

Some special flavors in foods are unpopular. Although a degree of bitterness is expected in beverages such as coffee, beer, and wine, a bitter taste is the main reason for the rejection of various food products. The degree of bitterness has proved a major limitation in the acceptance of commercial citrus juices. Two classes of chemical compounds, flavonoids (mainly naringin), and limonoids (mainly limonin) are largely responsible and have notable differences. A commercial process is therefore needed that removes bitter components without adding or changing food products, while maintaining the unique and expected flavor and the nutritional value of the products [9].

To suppress unpleasant odors or tastes, CDs can be used to remove or mask undesirable components. Some foods have a peculiar smell, but when CDs are added during manufacturing, these components can form CD inclusion complexes, deodorizing the resulting products. With the aim of masking the bitter taste using CDs, Binello and colleagues prepared a series of macromolecular derivatives in which

β-CD and γ-CD were covalently bound to carboxy methyl chitosan (CM-chitosan) and carboxymethyl cellulose (CM-cellulose), which worked successfully on limonin, naringin, and caffeine [11].

1.2.3 *Applications in protecting food ingredients*

The formation of inclusion complexes with CDs can protect some food components that are sensitive to degradation by oxygen, heat, or light. In the production of juices, the mechanical damage suffered by vegetable and fruit tissues often leads to rapid enzyme-catalyzed browning reactions. Polyphenoloxidase converts colorless polyphenols into colored compounds, and fruit and vegetable juices can be treated with CDs to remove this enzyme by complexation [12].

1.3 Applications in environmental protection

1.3.1 *Applications in wastewater treatment*

Water availability has become a major problem in most regions of the world due to a dearth of water resources, population explosion, industrial pollution, irrational consumption, and climate change. The demand for water purification to ensure safe, usable water has therefore increased [13, 14].

Superparamagnetic iron-oxide nanoparticles (SPION) and their nanocomposites with β-cyclodextrins (SPION/β-CDs) have been synthesized that are capable of scavenging oil from water, as demonstrated by oil sorption and magnetic separation. An oil retention capacity of 7.2 g/g of nanocomposite was achieved. In addition, 94.22% degradation of malachite green dye was achieved in a 2 h sunlight-driven photocatalysis process. The high removal efficiency and reusability confirms the suitability of such materials for oil spill remediation [14].

1.3.2 *Applications in the remediation of pesticide-contaminated soils*

Soil contamination of agricultural land is a major issue, but it is expensive to rectify and is not receiving the attention it deserves.

For example, alternative low-cost and easily implementable solutions are needed to accelerate the degradation and natural attenuation of pesticides [15]. Sorptive interactions between the soil matrix and organic contaminants are dependent of the physicochemical characteristics of soil/chemical combinations.

CDs have a low-polarity cavity within which organic compounds of an appropriate shape and size can form inclusion complexes. This property provides CDs with the capacity to increase the apparent solubility of some hydrophobic pollutants such as chlorinated solvents, pesticides, and nitroaromatic compounds.

Due to their nontoxic and biodegradable properties, CDs have been applied to promote the degradation of hazardous pollutants discharged into aqueous environments. Furthermore, CDs present several advantages over traditional organic solvents and nonionic surfactants, including lower toxicity and higher biodegradability [16]. Jaime and colleagues tested β-CDs in the remediation of pesticide-contaminated soils and demonstrated the formation of stable inclusion complexes containing the herbicide norflurazon (NFL) as a guest with a β-CD host [16, 17].

1.4 Applications in pharmaceuticals

1.4.1 *Applications in controlled drug release*

Polysaccharide gold nanocluster supramolecular conjugates containing cyclodextrin-modified hyaluronic acid (HACD) and gold nanoparticles bearing adamantane moieties (AuNPs) have a porous structure, as shown in Fig. 1.4 [7]. These molecules have great potential for the targeted delivery of anticancer drugs. The HACD–AuNPs supramolecular conjugates could serve as a platform for the controlled delivery of various anticancer drugs including doxorubicin hydrochloride (DOX), paclitaxel (PTX), camptothecin (CPT), irinotecan hydrochloride (CPT-11), and topotecan hydrochloride (TPT) by controlling the association/dissociation of drug molecules from the cavities formed by HACD and AuNPs, facilitating the efficient targeting of cancer cells via hyaluronic acid (HA). Drug–HACD–AuNPs

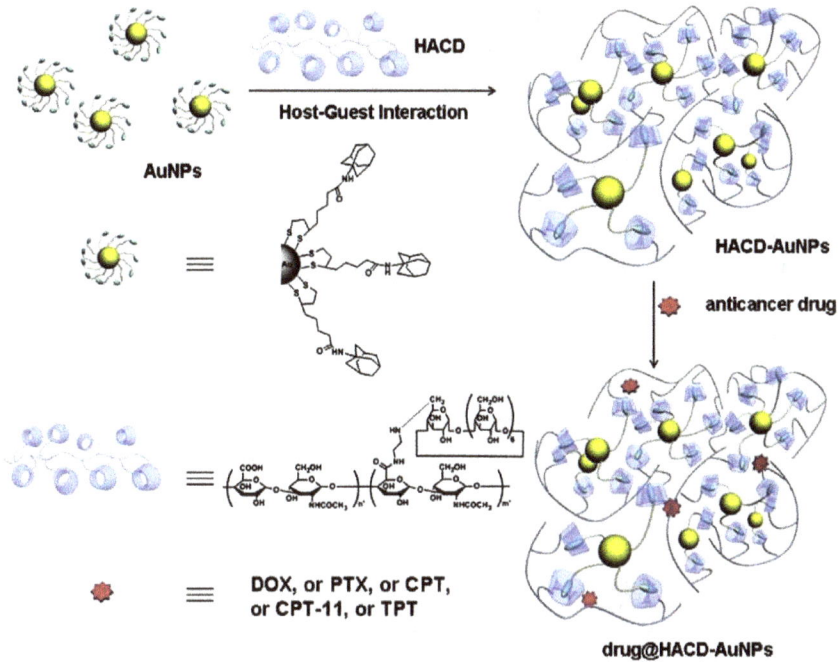

Fig. 1.4. Schematic illustration of the chemical structures and construction of HACD–AuNPs and drug–HACD–AuNPs complexes [7]. HACD, CD-modified hyaluronic acid; AuNPs, gold nanoparticles bearing adamantane moieties; DOX, doxorubicin hydrochloride; PTX, paclitaxel; CPT, camptothecin; CPT-11, irinotecan hydrochloride; TPT, topotecan hydrochloride.

systems that are responsive to pH could efficiently release the cargo (drug) into the mildly acidic environment of cancer cells.

Ethem and colleagues synthesized two comb-like grafting copolymers through free-radical copolymerization between two types of methyl acrylate monomers modified with a CD and adamantane (Ada), respectively [18]. Due to the inclusion complexation of the side groups (CDs and Ada), the viscosity of the system increased substantially upon gel formation upon mixing the two polymer solutions at a certain concentration. Poly (organophosphazene) structures based on noncovalent inclusion complexation were also used to prepare supramolecular gels that respond to stimuli [19]. Additionally, a minimally invasive injectable shear-thinning HA hydrogel [20] was developed through host–guest interactions between Ada-modified HA (guest macromer) and β-CD-modified HA (host macromer). The host–guest assembly mechanism made the injection easy due to the shear-thinning behavior and facilitated the rapid reassembly needed for material retention at the target. The erosion and release of a biomolecule model of this hydrogel were investigated in detail, and the release was sustained for more than 60 days.

As illustrated in Fig. 1.5, two biocompatible homopolymers, PEG-β-CD, polyethylene glycol (PEG) modified with β-CD, and PLLA-Fc, poly L-lactide (PLLA) modified with ferrocene (Fc), were used to prepare electrochemical-responsive micelles that can control drug release [21, 22]. A series of cyclodextrin-based star polymers were efficiently synthesized using β-CD as a hydrophilic core and methyl methacrylate (MMA) and tert-butyl acrylate (tBA) as hydrophobic arms.

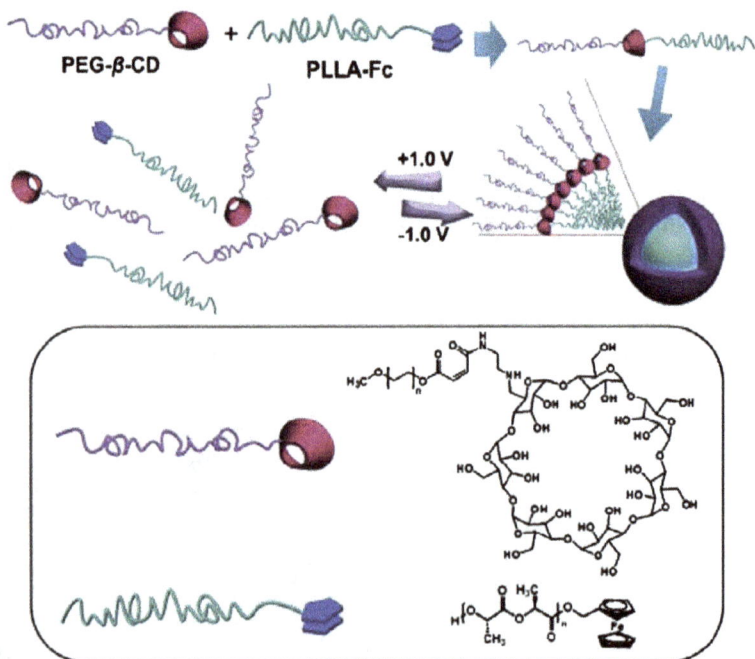

Fig. 1.5.   Structures of PEG-β-CD and PLLA-Fc and a schematic of the voltage-responsive assembly and disassembly of PLLA-Fc/PEG-β-CD micelles [22]. PEG, poly ethylene glycol; PLLA, poly L-lactide; Fc, ferrocene.

MMA–tBA–CD copolymers are capable of forming nanoparticles <200 nm in diameter [23]. Fluorescently-labeled CD-based nanoparticles (CD-NPs) were readily internalized by adenocarcinomic human alveolar basal epithelial cells (A549 cells) relative to heterogeneous human epithelial colorectal adenocarcinoma cells (Caco-2 cells) in a time-dependent manner. Particles showed preferential, concentration-dependent accumulation in the cytoplasm with superior uptake of positively charged particles over negatively charged ones. CD-NPs loaded with the antineoplastic idarubic ensured a sustained release of the payload over 48 h. These results suggest CD-star copolymers are a versatile new class of nanochemotherapeutics. However, the biodegradability, biocompatibility, uptake mechanisms, and *in vivo* efficacy of these copolymers remains to be investigated [23].

The supramolecular interaction of gemifloxacin and hydroxy-propyl-β-cyclodextrin (HP-β-CD), and the supramolecular interaction of moxifloxacin and β-CD were studied to establish a method for the analysis of gemifloxacin and moxifloxacin [24, 25]. HP-β-CD reacted with gemifloxacin to form a 1:1 host–guest inclusion complex, and β-CD reacted with moxifloxacin to form a 1:1 host–guest inclusion complex. A linear relationship between the intensity and gemifloxacin concentration was observed across a concentration range of 20–140 ng/mL with a high correlation coefficient (0.9997) and 10–60 ng/mL for moxifloxacin (correlation coefficient = 0.9997). The method was successfully applied to the analysis of gemifloxacin and moxifloxacin for pharmaceutical preparation [24, 25].

In order to obtain a controlled-drug-release textile material for patients with skin diseases, a soft, hydrophilic, and aseptic textile free from pathogenic agents is needed and was achieved. Using these conditions, the active CD derivative monochlorotriazinyl (MCT)-β-CD can be grafted onto the textile material [26]. Based on the therapeutic dose indicated, the active medicine is applied in optimal complexing conditions (low temperature and long duration), and drug release is then assessed *in vitro* using a perspiration kit at 37°C (physiological temperature). The *in vitro* release results must be identical in terms of drug release duration and diffusion flow (mg/24 h) to those indicated by the therapeutic dose. Such procedures must be carried out with materials that have been professionally manufactured and clinically tested [26].

Reagents are used to fix CD derivatives onto fibrous polymers by bridging between the nucleophilic groups of the fibrous polymers and CDs. Suitable reaction conditions for grafting CDs onto appropriate textiles are as follows:

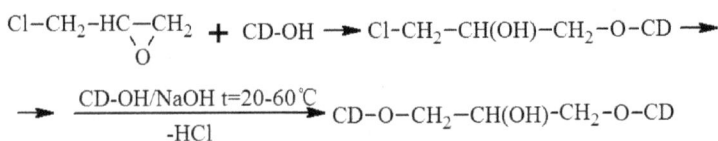

$$Cl-CH_2-HC-CH_2 \; (O) \; + \; CD\text{-}OH \longrightarrow Cl-CH_2-CH(OH)-CH_2-O-CD \longrightarrow$$

$$\longrightarrow \xrightarrow[-HCl]{CD\text{-}OH/NaOH \; t=20\text{-}60\,^{\circ}C} CD-O-CH_2-CH(OH)\text{-}CH_2-O-CD$$

Reaction with epichlorohydrin (EPCL).

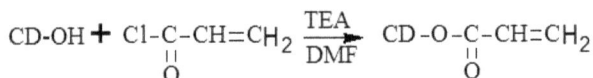

$$CD\text{-}OH + Cl-C-CH=CH_2 \; \overset{\displaystyle\|}{O} \; \xrightarrow[DMF]{TEA} \; CD-O-C-CH=CH_2 \; \overset{\displaystyle\|}{O}$$

Reaction with acryloyl derivatives. TEA, triethylamine; DMF, dimethyl formamide.

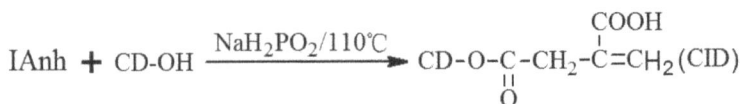

$$IAnh \; + \; CD\text{-}OH \; \xrightarrow{NaH_2PO_2/110^{\circ}C} \; CD-O-C-CH_2-\overset{\displaystyle COOH}{\underset{\displaystyle\|}{\underset{O}{C}}}=CH_2 \, (CID)$$

Reaction with itaconic anhydride (IAnh). CID, cyclodextrinitaconate derivatives.

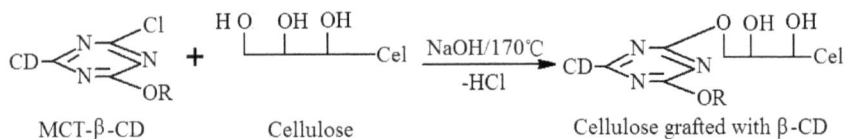

$$CD-\!\!\!<\!\!\!\begin{smallmatrix}N\\N\end{smallmatrix}\!\!\!>\!\!\!\begin{smallmatrix}Cl\\N\\OR\end{smallmatrix} + \overset{H\;O\;\;OH\;OH}{\underset{Cel}{| \; | \quad | \quad |}} \; \xrightarrow[-HCl]{NaOH/170^{\circ}C} \; CD-\!\!\!<\!\!\!\begin{smallmatrix}N\\N\end{smallmatrix}\!\!\!>\!\!\!\begin{smallmatrix}O\;OH\;OH\\N\\OR\end{smallmatrix}\underset{Cel}{| \quad | \quad |}$$

MCT-β-CD   Cellulose   Cellulose grafted with β-CD

Reaction of monochlorotriazinyl-β-cyclodextrin (MCTβ-CD) with cellulose.

Depending on the actual drug release conditions, model liquids comprising chemical species and concentrations similar to the biological conditions studied must be chosen. Table 1.1 shows drug release values for four drugs with two hosts tested in different release conditions.

Table 1.1. Drug release values from CDs.

Host (shape)	Drug	Release conditions	Amount of drug $M^{\infty a}$ (mg/g)	Release time (h)	Ref.
β-CD + EPCL[b] hydrogel (disk)	Naproxen (anti-inflammatory)	pH 7.0, 37°C	11.5	30	[27]
β-CD + EPCL hydrogel (disk)	Naproxen	pH 1.2, 37°C	1.5	30	[27]
β-CD + EPCL hydrogel (disk)	Naftifine (antifungal)	pH 1.2, 37°C	1.4	30	[27]
β-CD + EPCL hydrogel (disk)	Terbinafine (antifungal)	pH 1.2, 37°C	1.7	30	[27]
PET + β-CD + CTR[c]	Ciprofloxacin (antibiotic)	Water	62	1800	[28, 29]
PET + β-CD + CTR	Ciprofloxacin	Buffer, pH 7.4	45	6	[28, 29]
PET + β-CD + CTR	Ciprofloxacin	Human plasma	52	24	[28, 29]
PET + β-CD + CTR	Ciprofloxacin	Pure water	66	1320	[28, 29]
PET + β-CD + CTR	Ciprofloxacin	pH 2, 37°C	61	7	[28, 29]

Note: [a] M^{∞} was calculated using the Korsmeyer–Peppas equation.
[b] EPC, epichlorohydrin.
[c] PET, polyester; CTR, citric acid.

Information from studies on medical textile fabrics for antibacterial therapies is listed in Table 1.2. Release systems for antibiotics, including antibacterial active principles, have been improved [26].

Due to their supramolecular characteristics, CDs can form a drug-release system that improves the solubility of lipophilic drugs. Efficient inclusion requires a lipophilic drug with appropriate molecular dimensions to allow it to fit the inner cavity of CDs. The drug is then released following physiological cutaneous stimuli upon coming into contact with sweat, either from the hydrophobic cavity or through the biodegradation of biocompatible layers when the system is attached to a textile material [26].

The field of CDs is rapidly expanding, as evidenced by the increase in scientific papers on the subject, and significant advances are likely to be made in the near future.

1.4.2 *Applications in protecting active ingredients*

Some drugs and drug candidates are unstable and easily damaged by oxygen, heat, light, or enzymes. Since CDs can interact with drugs to increase their solubility and stability, they have been used extensively in pharmaceutical research and development. Doxycycline (Doxy) belongs to the tetracycline group of broad-spectrum antibiotics. Doxy has low stability in aqueous solution, which restricts its ophthalmic clinical application. The stability of Doxy/HP-β-CD complex in both solid and aqueous states was evaluated. The results showed that the formation of an inclusion complex with HP-β-CD improved the situation, and the Doxy/HP-β-CD complex has good *in vitro* antibacterial activity and holds promise for clinical applications [30].

1.5 Applications in biological fields

1.5.1 *Applications as gene carriers*

Hyperbranched polymers have attracted extensive interest in recent years. They are examples of dendritic polymers with broad industrial applications as solid carriers of catalysts, biological materials, and drugs [3]. Ritter and coworkers combined the host–guest recognition

Table 1.2.   Antibacterial medical textiles with CDs [26].

Textile	Active principle/Drug	Host reagent	Biological tests
Wool	Ag (NPs)[a], triclosan	MCT-β-CD[b]	After 15 washing cycles, the antibacterial efficiency exceeds 75%
Cotton	Octenidinedi-hydrochloride	β-CD grafted with BTCA[c]	Two types of bacteria and fungi/Diffusion Disk Method/over 20 washing cycles
Cotton	Silver (I)	CD grafted with CTR[d]	*Escherichia coli*/7 days of observation
Cotton	AgNO3 reduced with NaBH4 or β-CD-g-PAA[e]	MCT-β-CD-g-PAA	*Escherichia coli/Staphylococcus aureus*
Cotton	Miconazole nitrate (antimycosis)	MCT-β-CD	Tests on *Candida albicans, aurococcus* and *colon bacillus*
Cotton	Ag (NPs)	MCT-β-CD	*Staphylococcus aureus, Escherichia coli*
Cationized cotton	Colloid Ag (NPs)	MCT-β-CD-g-PAA with EPCL/β-CD-g-PAA with EPCL[f]	G +ve and G −ve bacteria
Hemp fibers	Ferulic acid, caffeic acid, ethyl ferulate allantoin	MCT-β-CD	Microbiological analysis
PET[g] vascular prostheses	Ciprofloxacin (antibiotics)	HP-β-CD, HP-γ-CD, Me-β-CD[h]	*Staphylococcus aureus/Escherichia coli*

[a]NPs, silver nanoparticles.
[b]MCT, monochlorotriazinyl.
[c]BTCA, butanetetracarboxylic acid.
[d]CTR, citric acid.
[e]PAA, polyacrylic acid.
[f]EPCL, epichlorohydrin.
[g]PET, polyester.
[h]HP, hydroxypropyl; Me, methyl.

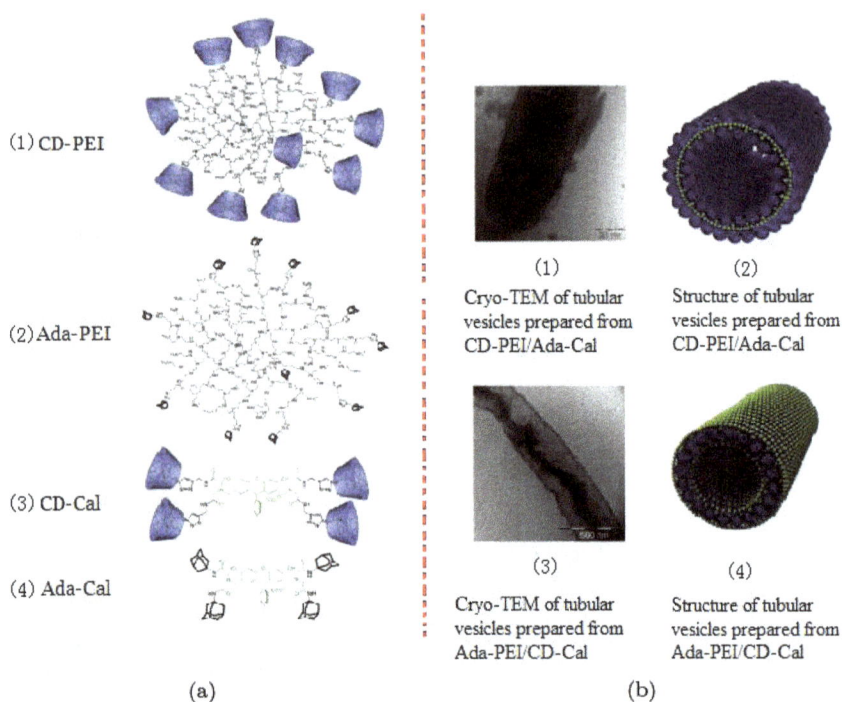

Fig. 1.6.   (a): Structures of CD-PEI, Ada-PEI, CD-Cal, and Ada-Cal; (b): Cryo-TEM of tubular vesicles prepared from CD-PEI/Ada-Cal and Ada-PEI/CD-Cal [31]. PEI, polyethylenimine; Cal, calcein; Ada, adamantine; TEM, transmission electron microscopy.

of CD and Ada (adamantane) with hyperbranched polyethylenimine (PEI) and synthesized hyperbranched PEI tubular micelles for the first time [31]. β-CD and Ada were connected to both the hyperbranched PEI and the fluorescent dye calcein through efficient click chemistry (Fig. 1.6(a)), and the products assembled to form nanotubular vesicles in the aqueous phase via the inclusion complexation of CD and Ada, as confirmed by 2D rotating frame overhauser effect spectroscopy (ROESY). Due to the hydrophilic properties of the outer wall of CDs, different fluorescence effects can be obtained using diverse assembly modes (Fig. 1.6(b)). PEI is regularly used as a skeleton by various synthetic enzymes. Its surface was abundant positive charges and can adsorb negatively charged DNA. These advantageous properties make hyperbranched PEI a promising candidate gene and drug carrier [32].

1.5.2 *Applications in producing functionalized materials*

A stimuli-responsive microgel system tuned by ultraviolet (UV) light based on host–guest interactions has been prepared [33–35]. Characterization of the photoreversibility of the resulting microgel was performed by transmission electron microscopy (TEM), UV–Vis spectroscopy, and ^{1}H nuclear magnetic resonance (NMR) measurements. The microgel displayed reversible size changes upon exposure to UV–Vis irradiation. Supramolecular systems with a wide range of light responsiveness that undergo macroscopic self-assembly based on CD/Azo (azobenzene) host–guest molecular recognition have also been described [33–35]. Such photoswitchable self-assembly and photoresponsive properties can be used to produce functionalized materials, which are expected to have wide applicability in a variety of fields, including artificial muscles on the macroscopic scale. Human recombinant single chain fragment variable (scFv) antibodies for site-specific photocoupling have been tailored using unnatural amino acid (UAA) and dock "n" flash technologies [36]. Specifically, the photoreactive UAA, *p*-benzoyl-L-phenylalanine (*p*Bpa), has been successfully introduced. A mutated scFv antibody was expressed in *E. coli* with retention of structural and functional properties, including binding affinity, and specifically photocoupled to free and surface (array) grafted *β*-CD after inclusion-complex formation. The mutated *p*Bpa-scFv (guest) and *β*-CD (host) formed an inclusion complex with moderate affinity, and NMR analysis indicated that the entire benzophenone ring might enter the cavity. In the subsequent photocoupling step, *β*-CD provides the hydrogen atoms to be abstracted [36].

1.5.3 *Applications in supramolecular architecture*

Over the last century, there has been a growing interest in natural porphyrinoids [4], which play a key role in the structures responsible for various vital functions, including oxygen binding (heme in hemoglobin and myoglobin), the electron transport chain (cytochrome oxidase), metabolism (cytochrome P450), and photosynthesis (chlorophyll). Supramolecular structures consisting of metal holoporphyrins display unique properties that are inaccessible by other means. Therefore, synthetic porphyrinoids have been widely examined as artificial enzymes, especially as functional analogs of cytochrome oxidase, myoglobin, and hemoglobin. Moreover, porphyrinoids are useful as photo sensitizers in photodynamic therapy (PDT), which is a promising way to treat cancer. Porphyrinoid-CD systems are constructed by both covalent linking and the formation of inclusion complexes. They have been investigated as biomimetics that imitate natural carotenoid cleavage, cytochrome P450-mediated hydroxylation, and oxygen binding by hemoglobin.

Moreover, porphyrinoids can mimic the functions of natural structures in light-harvesting systems. Supramolecular assemblies of these components are of particular interest in this field, as they can provide a facile route to multichromophoric arrays. In addition, porphyrin and phthalocyaninemacrocyclic rings can be considered as artificial light-harvesting antennae, in which the CD counterpart of the system provides the desired supramolecular architecture. CDs improve the photosensitizing properties of porphyrinoids by increasing their singlet oxygen quantum yield. This is of immense value for PDT [4].

1.6 Applications in other fields

1.6.1 *Applications in cosmetics*

Exposure to UV radiation from the sun (290−400 nm) is associated with several harmful effects on human skin including erythema, photoaging, immune suppression, and skin-cutaneous cancer, the latter representing the most prevalent form of human neoplasia. These findings have prompted the widespread use of topical sun protective preparations [37].

Since 4-methylbenzylidene camphor (4-MBC) is an important organic UV filter, it is widely used in sunscreen and cosmetic products.

CDs can entrap appropriately sized lipophilic compounds into their hydrophobic cavities, which can enhance their stability to air and light. A study demonstrated that complexation of 4-MBC with random methyl-β-cyclodextrin (RM-β-CD) markedly reduced the degradation of the sunscreen agent without affecting its localized distribution on the skin surface. The 4-MBC/RM-β-CD complex therefore displays improved UV filter efficacy [38].

1.6.2 *Applications in textiles*

CDs can form inclusion complexes with a large number of organic molecules, and this property enables them to be used in a variety of different textile applications. Since CDs can incorporate different dyes into their cavity, they should be able to act as retarders in a dyeing process.

Cationic dyes have very low migration power on polyacrylonitrile (PAN) fibers due to their high substantivity and rapid uptake over a small temperature range above the T_g of the fiber. Color intensity can be improved by the use of different retarding reagents. β-CD was investigated in the dyeing of PAN fibers with cationic dyes, and effective dyeing was improved when β-CD was used as a retarding reagent, compared with the cationic retarding reagent [39].

1.6.3 *Applications in analysis*

Some polymeric systems based on CDs and various guest molecules are sensitive to changes in temperature, illumination intensity, humidity, and electrical signal, and their structures are changed when conditions change.

A supramolecular triblock stimuli-responsive copolymer, poly (N-isopropylacrylamide)-block-poly (e-caprolactone)-block-poly-(N,N-dimethyl aminoethyl methacrylate) (PNIPAM-*b*-PCL-*b*-PDMAE MA), was developed by thiolene Michael-type addition and host–guest interaction (Fig. 1.7) [40]. The three blocks PNIPAM, PCL, and PDMAEMA were synthesized via reversible addition fragmentation chain transfer polymerization (RAFT), ring opening polymerization

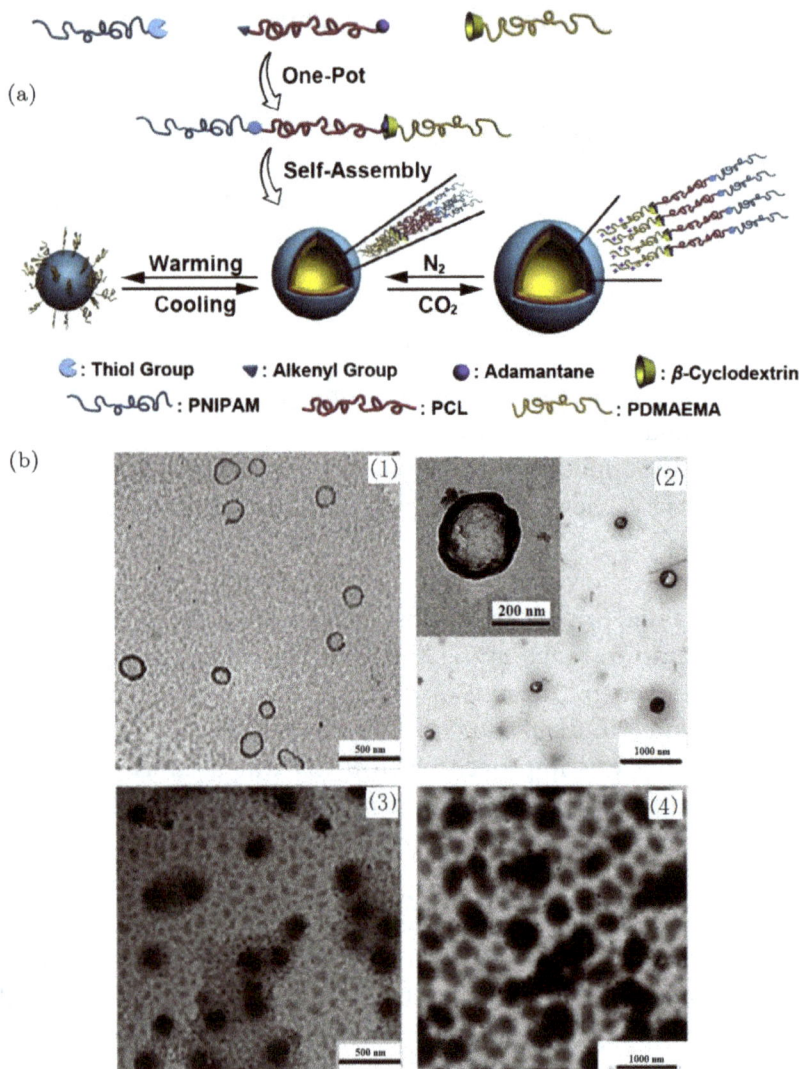

Fig. 1.7.   (a) Illustration of the synthesis and self-assembly of the supramolecular triblock copolymer PNIPAM-*b*-PCL-*b*-PDMAEMA andits response to CO_2 and temperature; (b) TEM images of PNIPAM-*b*-PCL-*b*-PDMAEMA self-assembling aggregates in water before (1) and after (2) treatment with CO_2 at 2°C. Samples used in (3) and (4) were prepared at 40 and 80°C, respectively [40]. PNIPAM, poly N-isopropylacrylamide; PCL, poly e-caprolactone; PDMAEMA, poly (*N,N*)-dimethyl aminoethyl methacrylate.

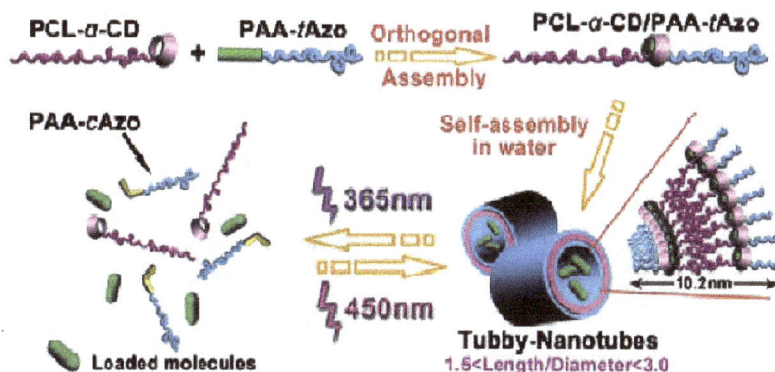

(a)

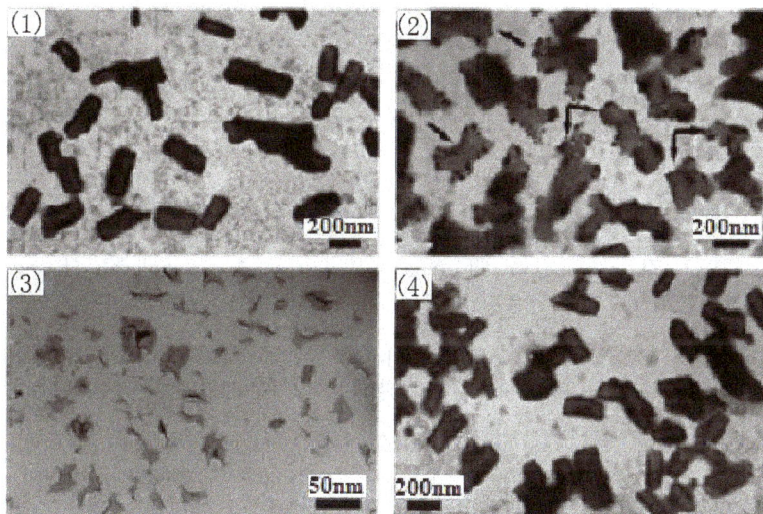

(b)

Fig. 1.8. (a) Chemical structures of PCL-α-CD and PAA-tAzo, and a schematic representation of the reversible assembly and disassembly of one-dimensional light-responsive nanotubes; (b) TEM images of the reversible assembly and disassembly of PCL-α-CD/PAA-tAzo nanotubes [41]. PCL, poly e-caprolactone; PAA, polyacrylic acid; Azo, azobenzene.

(ROP), and atom transfer radical polymerization (ATRP), respectively. The triblock copolymer PNIPAM-*b*-PCL-*b*-PDMAEMA could self-assemble into vesicles, and it displayed reversible variation in morphology and size in response to CO_2 and temperature.

Polymeric systems based on CD/Azo host–guest linkers can reversibly assemble and disassemble under external photoirradiation of alternate UV/visible light due to the isomerization of the Azo constituent. Two homopolymers, poly (caprolactone)-α-CD (PCL-α-CD) and poly (acrylic acid)-trans-Azo (PAA-tAzo), were designed and synthesized and can form one-dimensional nanotubes in water as shown in Fig. 1.8 [41]. The reversible assembly and disassembly behavior of PCL-α-CD/PAA-tAzo nanotubes was visualized by TEM. As expected, these supramolecular nanotubes can be used in the controllable release of fluorescent rhodamine B (RB), and the release speed could be precisely controlled by adjusting the duration of irradiation.

References

1. Ogawa N, Takahashi C, Yamamoto H. (2015). Physicochemical characterization of cyclodextrin-drug interactions in the solid state and the effect of water on these interactions. *Journal of Pharmaceutical Sciences*, 104, pp. 942–954.
2. Gharib R, Greige-Gerges H, Fourmentin S, Charcosset C, Auezova L. (2015). Liposomes incorporating cyclodextrin-drug inclusion complexes: current state of knowledge. *Carbohydrate Polymers*, 129, pp. 175–186.
3. Liu BW, Zhou H, Zhou ST, Yuan JY. (2015). Macromolecules based on recognition between cyclodextrin and guest molecules: synthesis, properties and functions. *European Polymer Journal*, 65, pp. 63–81.
4. Kryjewski M, Goslinski T, Mielcarek J. (2015). Functionality stored in the structures of cyclodextrin–porphyrinoid systems. *Coordination Chemistry Reviews*, 300, pp. 101–120.
5. Yue X, Chen X, Li Q, Qian Z. (2015). Soft aggregates formed by a nonionic phytosterol ethoxylate and β-cyclodextrin in aqueous solution. *Colloids and Surfaces A: Physicochemical and Engineering Aspects*, 482, pp. 79–86.
6. Manakker FVD, Vermonden T, Nostrum CFV, Hennink WE. (2009). Cyclodextrin-based polymeric materials: synthesis, properties, and pharmaceutical/biomedical applications. *Biomacromolecules*, 10(12), pp. 57–75.
7. Li N, Chen Y, Zhang YM, Yang Y, Su Y. (2014). Polysaccharide gold nanocluster supramolecular conjugates as a versatile platform for the targeted delivery of anticancer drugs. *Scientific Reports*, 4(41), pp. 64–70.
8. Grimmo AEP, Lee SK. (1998). Retention and release of aroma compounds in foods containing proteins. *Food Technology*, 52(5), pp. 68–74.
9. Astray G, Gonzalezbarreiro C, Mejuto JC, Rialotero R, Simalgándara J. (2009). A review on the use of cyclodextrins in foods. *Food Hydrocolloids*, 23(7), pp. 1631–1640.
10. Jouquand C, Ducruet V, Giampaoli P. (2004). Partition coefficients of aroma compounds in polysaccharide solutions by the phase ratio variation method. *Food Chemistry*, 85(85), pp. 467–474.

11. Binello A, Cravotto G, Nano GM, Spagliardi P. (2004). Synthesis of chitosan-cyclodextrin adducts and evaluation of their bitter-masking properties. *Flavor & Fragrance Journal*, 19(19), pp. 394–400.
12. Valle EMD. Cyclodextrins and their uses: a review. (2004). *Process Biochemistry*, 39(9), pp. 1033–1046.
13. Ahmed S, Rasul MG, Brown R, Hashib MA. (2011). Influence of parameters on the heterogeneous photocatalytic degradation of pesticides and phenolic contaminants in wastewater: a short review. *Journal of Environmental Management*, 92(3), pp. 311–330.
14. Kumar A, Sharma G, Mu N, Thakur S. (2015). SPION/β-cyclodextrin core-shell nanostructures for oil spill remediation and organic pollutant removal from waste water. *Chemical Engineering Journal*, 280, pp. 175–187.
15. Shea PJ, Machacek TA, Comfort SD. (2004). Accelerated remediation of pesticide-contaminated soil with zerovalent iron. *Environmental Pollution*, 132(2), pp. 183–188.
16. Zeng QR, Tang HX, Liao BH, Zhong T, Tang C. (2006). Solubilization and desorption of methyl-parathion from porous media: a comparison of hydroxypropyl-beta-cyclodextrin and two nonionic surfactants. *Water Research*, 40(7), pp. 1351–1358.
17. Villaverde J. (2007). Time-dependent sorption of norflurazon in four different soils: use of β-cyclodextrin solutions for remediation of pesticide-contaminated soils. *Journal of Hazardous Materials*, 142, pp. 184–190.
18. Kaya E, Mathias LJ. (2010). Synthesis and characterization of physical crosslinking systems based on cyclodextrin inclusion/host-guest complexation. *Journal of Polymer Science A: Polymer Chemistry*, 48(3), pp. 81–92.
19. Tian Z, Chen C, Allcock HR. (2014). Synthesis and assembly of novel poly (organophosphazene) structures based on noncovalent "host-guest" inclusion complexation. *Macromolecules*, 47(3), pp. 1065–1072.
20. Rodell CB, Kaminski AL, Burdick JA. (2013). Rational design of network properties in guest-host assembled and shear-thinning hyaluronic acid hydrogels. *Biomacromolecules*, 14(11), pp. 25–34.
21. Tomatsu I, Hashidzume A, Harada A. (2006). Contrast viscosity changes upon photoirradiation for mixtures of poly (acrylic acid)-based α-cyclodextrin and azobenzene polymers. *Journal of the American Chemical Society*, 128(222), pp. 6–7.
22. Peng L, Feng AC, Zhang H, Wang H, Jian CM, Liu BW. (2014). Voltage responsive micelles based on assembly of two biocompatible homopolymers. *Polymer Chemistry*, 5(5), pp. 1751–1759.
23. Nafee N, Hirosue M, Loretz B, Wenz G, Lehr CM. (2015). Cyclodextrin-based star polymers as a versatile platform for nanochemo therapeutics: enhanced entrapment and uptake of idarubicin. *Colloids and Surfaces B: Biointerfaces*, 129, pp. 30–38.
24. Dsugi NFA, Elbashir AA. (2015). Supramolecular interaction of gemifloxacin and hydroxyl propyl β-cyclodextrin spectroscopic characterization, molecular modeling and analytical application. *Spectrochimica Acta A: Molecular and Biomolecular Spectroscopy*, 151, pp. 360–367.

25. Dsugi NFA, Elbashir AA. (2015). Supramolecular interaction of moxifloxacin and β-cyclodextrin spectroscopic characterization and analytical application. *Spectrochimica Acta A: Molecular and Biomolecular Spectroscopy*, 137, pp. 804–809.
26. Radu CD, Parteni O, Ochiuz L. (2016). Applications of cyclodextrins in medical textiles-review. *Journal of Controlled Release*, 224, pp. 146–157.
27. Machín R, Isasi JR, Vélaz I. (2012). β-cyclodextrin hydrogels as potential drug delivery systems. *Carbohydrate Polymers*, 87, pp. 2024–2030.
28. Blanchemain N, Karrout Y, Tabary N, Neut C, Bria M, Siepmann J. (2011). Methyl-β-cyclodextrin modified vascular prosthesis: influence of the modification level on the drug delivery properties in different media. *Acta Biomaterialia*, 7, pp. 304–314.
29. Blanchemain N, Karrout Y, Tabary N, Bria M, Neut C, Hildebrand HF, Siepmann J. (2012). Comparative study of vascular prostheses coated with poly-cyclodextrins for controlled ciprofloxacin release. *Carbohydrate Polymers*, 90, pp. 1695–1703.
30. Zhang H, Chen M, He Z. (2013). Molecular modeling-based inclusion mechanism and stability studies of doxycycline and hydroxypropyl-β-cyclodextrin complex for ophthalmic delivery. *American Association of Pharmaceutical Scientists*, 14(1), pp. 10–18.
31. Bohm I, Isenbugel K, Ritter H, Branscheid R, Kolb U. (2011). Cyclodextrin and adamantane host-guest interactions of modified hyper-branched poly (ethylene imine) as mimetics for biological membranes. *Angewandte Chemie International Edition*, 50(34), pp. 6–9.
32. Jin H, Liu Y, Zheng Y, Huang W, Zhou Y, Yan D. (2011). Cyto mimetic large scale vesicle aggregation and fusion based on host-guest interaction. *Langmuir*, 28(4), pp. 66–72.
33. Takashima Y, Nakayama T, Miyauchi M, Kawaguchi Y, Yamaguchi H, Harada A. (2004). Complex formation and gelation between copolymers containing pendant azobenzene groups and cyclodextrinpolymers. *Chemical Letters*, 33(7), pp. 890–891.
34. Tomatsu I, Hashidzume A, Harada A. (2005). Photoresponsive hydrogel system using molecular recognition of α-cyclodextrin. *Macromolecules*, 38(522), pp. 3–7.
35. Yamaguchi H, Kobayashi Y, Kobayashi R, Takashima Y, Hashidzume A, Harada A. (2012). Photoswitchable gel assembly based on molecular recognition. *Nature Communications*, 3, p. 603.
36. Petersson L, Städe LW, Brofelth M, Gartner S, Fors E. (2014). Molecular design of recombinant scFv antibodies for site-specific photocoupling to β-cyclodextrin in solution and onto solid support. *Biochimicaet Biophysica Acta*, 1844, pp. 2164–2173.
37. Scalia S, Casolari A, Iaconinoto A, Simeoni S. (2002). Comparative studies of the influence of cyclodextrins on the stability of the sunscreen agent, 2-ethylhexyl-p-methoxycinnamate. *Journal of Pharmaceutical and Biomedical Analysis*, 30(4), pp. 1181–1190.

38. Scalia S, Tursilli R, Iannuccelli V. (2007). Complexation of the sunscreen agent, 4-methylbenzylidene camphor with cyclodextrins: effect on photostability and human stratum corneum penetration. *Journal of Pharmaceutical and Biomedical Analysis*, 44(1), pp. 29–34.

39. Vončina B, Vivod V, Jaušovec D. (2007). β-Cyclodextrin as retarding reagent in polyacrylonitrile dyeing. *Dyes and Pigments*, 74(3), pp. 642–646.

40. Liu BW, Zhou H, Zhou ST, Zhang HJ, Feng AC, Jian CM. (2014). Synthesis and self-assembly of CO_2-temperature dual stimuli-responsive triblock copolymers. *Macromolecules*, 47(9), pp. 38–46.

41. Yan Q, Xin Y, Zhou R, Yin Y, Yuan J. (2011). Light-controlled smart nanotubes based on the orthogonal assembly of two homopolymers. *Chemical Communications*, 47(34), pp. 9594–9596.

2. General Methods for the Preparation of Cyclodextrin Inclusion Complexes

Jinpeng Wang, Haoran Fan,
and Mengke Zhang

*State Key Lab of Food Science and Technology,
School of Food Science and Technology,
Jiangnan University,
Wuxi 214122, China*

2.1 Introduction

Cyclodextrins (CDs) are widely used as host molecules for forming inclusion complexes with small hydrophobic guest molecules. In such inclusion complexes, guests occupy the hydrophobic cavity or gap between adjacent cyclodextrin molecules, and the properties of occluded guests are improved compared with those of free-guest molecules. For example, better solubility of inclusion complexes in solution is of benefit to food, agriculture, cosmetic, and pharmaceutical industries and to stockbreeding. Similarly, the greater stability of inclusion complexes prevents the degradation and oxidation of guests from heat and UV light irradiation, and inclusion complexes can often be stored in the environment for a longer time. Furthermore, unique properties can be bestowed on the guest, such as increased safety, lower stimulatory activity, and improved target delivery. Several methods are used to prepare inclusion complexes, including saturating solutions, spray-drying, cogrinding, freeze-drying, and various others.

Different target inclusion complexes can be prepared using different types of CD and different preparation methods.

2.2 Thermodynamic parameters for inclusion complex preparation

The energy of the host or guest is generated or absorbed when the two species interact. Thermodynamic parameters including the standard free-energy change (ΔG), the standard enthalpy change (ΔH), and the standard entropy change (ΔS) can be obtained from the temperature dependence of inclusion complex preparation. The Gibbs-free-energy change (ΔG) of the process is derived from the stability constant (K_S) and can be calculated using the following equation:

$$\Delta G = -RT \ln K_S, \tag{2.1}$$

where R is the universal gas constant (8.314 J mol^{-1} K^{-1}), T is the experimental operating temperature (K), and K_S is the stability constant of the inclusion process. Reaction enthalpies can similarly be determined from K_S obtained at different temperatures using the Van't Hoff equation. If two sets of data are available (i.e., two K_S values determined at two different temperatures T_1, T_2), then

$$\Delta H = \frac{RT_1 T_2 \ln\left(\frac{K_2}{K_1}\right)}{T_2 - T_1}, \tag{2.2}$$

$$\Delta G = \Delta H - T\Delta S. \tag{2.3}$$

The Van't Hoff equation is employed to calculate enthalpy (ΔH) and entropy (ΔS) changes in two inclusion complexes.

$$\ln K = -\frac{\Delta H}{RT} + \frac{\Delta S}{R} \tag{2.4}$$

Usually, complex formation is associated with a negative ΔH value, whereas ΔS can be positive or negative, suggesting that, depending on the guest molecule, several forces are involved in complex formation. Thermodynamic parameters for different cyclodextrins are listed in Table 2.1.

Table 2.1.   Common guests for native CDs and their respective stability constants.

Guest	CD	K_S (M^{-1})	ΔH (kJ mol^{-1})	ΔS (J mol^{-1} K^{-1})	Ref.
Benzoic acid	α	794.2	−40.2	−75.5	[1]
Acetic acid	α	9.1	−11.2	−18.8	[2]
2-[[4-(Dimethylamino)-phenyl]azo]pyridine	α	2455.4	−29.1	−33.56	[3]
1-Adamantane-carboxylate	β	1659.6	−20.3	−12.3	[4]
1,10-Decanediol	β	2238.1	−26.0	−23.5	[5]
Flurbiprofen	β	4466.5	−14.9	19.8	[6]
Imidazole	β	1.9	−16.3	−46.9	[7]
1-Naphthol	β	1230.2	−11.3	22.1	[8]
Anthracene	γ	223.8	−23.0	−32.2	[9]
Pyrene	γ	1122.0	−42.7	−84.8	[9]
Benzoic acid	HTM-α-CD		−68.6	−179.8	[10]
L-cysteine	MPD-α-CD	947.9	−73.3	−44.6	[11]
3-Nitroaniline	HTM-α-CD	1659.6	−48.1	−113.1	[10]
Cu^{2+}	DNH-β-CD	870.9	−18.4	−4.0	[12]
Hexane	Water-soluble CD		−28.2	−97.3	[13]
4-Nitrophenol	HDM-β-CD	128.9	−13.9	−3.9	[14]
L-alanine	MPD-γ-CD	1059.2	−43.9	−87.2	[11]

Note: HTM-α-CD, hexakis(2,3,6-tri-O-methyl)-α-cyclodextrin;
MPD-α-CD, mono-[6-(1-pyridinio)-6-deoxy]-α-cyclodextrin;
DNH-β-CD, 6-deoxy-6-(N-histamino)-β-cyclodextrin;
HDM-β-CD, heptakis(2,6-di-O-methyl)-β-cyclodextrin;
MPD-γ-CD, mono-[6-(1-pyridinio)-6-deoxy]-γ-cyclodextrin.

2.3 Inclusion mathematical models and parameters

No covalent bonds are broken or formed during the formation of the inclusion complex. The main driving force for complex formation is the release of enthalpy-rich water molecules from the cavity. These forces include electrostatic (ion–ion, ion–dipole, dipole–dipole, dipole-induced dipole, and higher order terms), van der Waals, hydrophobic, hydrogen bonding, charge transfer, and π–π stacking interactions, as well as steric effects.

The molar ratio of CD to guest molecules (G) is usually 1:1 for inclusion complexes formed in solution, and K_S, the stability

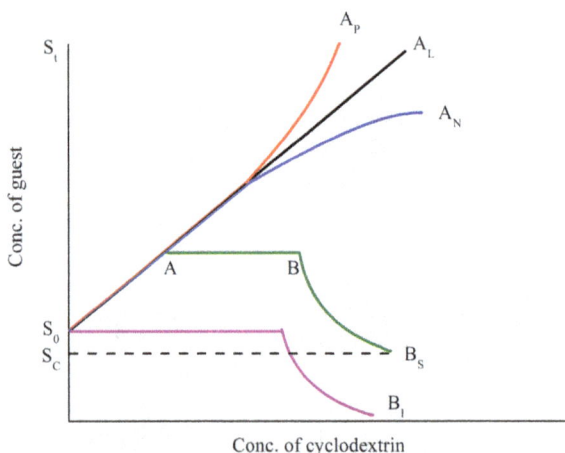

Fig. 2.1.   Phase solubility profiles and classification of complexes [16]. S_0 is the intrinsic solubility of the substrate (the dissolved guest/drug) in the aqueous complexation medium when no ligand (CD) is present.

(equilibrium) constant, represents the state change of the complex. For inclusion complexes with a host to guest molar ratio of 1:1, the equilibrium equation is as follows:

$$G + CD = GCD$$

$$K_S = \frac{[GCD]}{[G][CD]} \qquad (2.5)$$

From Eq. (5), K_S can be determined by measuring the concentration of the complex, guest, and cyclodextrin.

Higuchi and Connors [15] have classified complexes from their phase solubility profiles (Fig. 2.1) that are obtained from the interaction between the guest and the host (H) in solution. Phase solubility diagrams are classified into two main categories (A and B). A-type curves indicate the formation of soluble inclusion complexes, whereas B-type curves indicate the formation of inclusion complexes with poor solubility. A-type curves are subdivided into A_L-type (drug solubility increases linearly as a function of CD concentration), A_P-type (positively deviating isotherm), and A_N-type (negatively deviating isotherm) subtypes. B_I-type curves indicate the formation

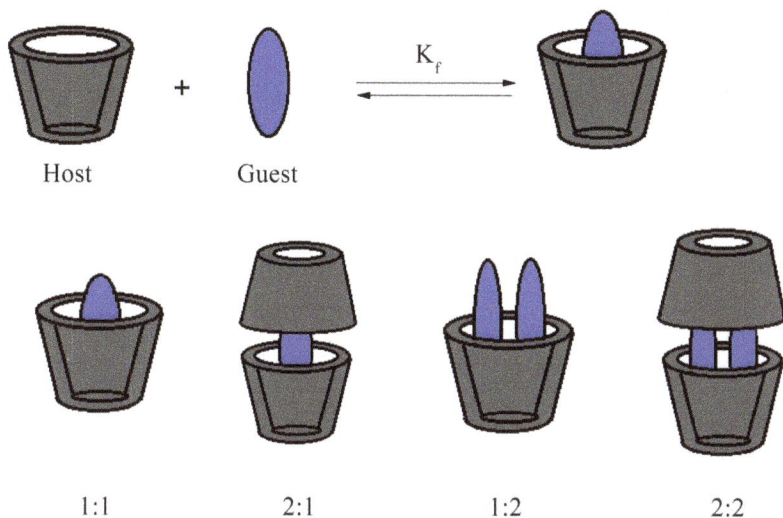

Fig. 2.2.   Different types of complexation.

of insoluble complexes, and B_S-type curves denote complexes with limited solubility. The negative deviation from linearity may be associated with ligand-induced changes in the dielectric constant of the solvent, or self-association of ligands at high CD concentrations.

β-CDs often gives rise to B-type curves due to their poor water solubility, whereas chemically modified CDs such as hydroxypropyl-β-cyclodextrin (HP-β-CD) and sulfobutyl ether-β-cyclodextrin (SBE-β-CD) usually produce soluble complexes and thus generate A-type curves. The A_L-type diagram is of first order with respect to the CD and may be of first or higher order with respect to the guest (G; i.e., GCD, G_2CD, G_3CD, ... , G_mCD). If the slope of an A_L-type system is greater than 1, higher-order complexes are indicated. A slope of less than 1 does not necessarily exclude higher-order complexation, but 1:1 complexation is usually assumed in the absence of other information. The A_P-type system suggests the formation of higher order complexes with respect to the CD at higher CD concentrations (i.e., GCD, GCD_2, GCD_3, ... , GCD_n). The stoichiometry of A_P-type systems can be evaluated by curve fitting, whereas the A_N-type system is problematic and difficult to interpret.

The most common cases are 1:1 CD:G complexes, except for complexes with long-chain or bifunctional guest molecules (e.g., guest molecules having two aromatic rings on opposite sides of a small central molecule segment), but 2:1 and 1:2 complexes have also been described (Fig. 2.2; e.g., 1-bromoadamantane:2 α-CD complexes or γ-CD:2 pyrene complexes). The formation of CD:G 1:2 or 2:1 complexes can be described using the following stability constants:

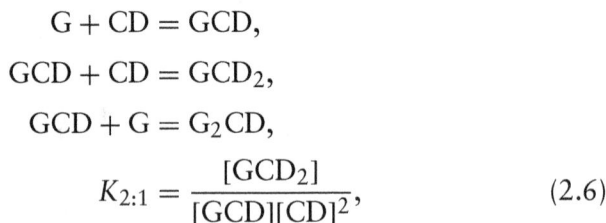

$$G + CD = GCD,$$

$$GCD + CD = GCD_2,$$

$$GCD + G = G_2CD,$$

$$K_{2:1} = \frac{[GCD_2]}{[GCD][CD]^2}, \tag{2.6}$$

$$K_{1:2} = \frac{[G_2CD]}{[G]^2[GCD]}. \tag{2.7}$$

In general, the formation of more complicated CD H:G complexes (G_mCD_n) can be described by the following equilibrium:

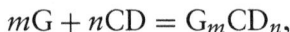

$$mG + nCD = G_mCD_n,$$

and the equation for the overall stability constant K_S:

$$K_S = \frac{[G_mCD_n]}{[G]^m[CD]^n}, \tag{2.8}$$

where

$$[G] = G_0,$$

$$[G]_t = G_0 + m[G_mCD_n],$$

$$[CD]_t = [CD] + n[G_mCD_n],$$

Therefore, values of [G_mCD_n], [G], and [CD] can be obtained from the following:

$$[G] = G_0,$$

$$[G_mCD_n] = \frac{[G]_t - G_0}{m},$$

$$[CD] = [CD]_t - n[G_mCD_n],$$

where G_0 is the intrinsic solubility of guest, $[G]_t$ is the total concentration of guest (complexed and uncomplexed), and $[CD]_t$ is the total concentration of CD. For phase solubility systems that are of first order with respect to the CD ($n = 1$), the following equation may be derived:

$$[G]_t = \frac{mKG_0^m[H]_t}{1 + KG_0^m} + G_0. \tag{2.9}$$

A plot of $[G]_t$ versus $[CD]_t$ for the formation of G_mCD_n should give a straight line with the y intercept representing G_0 and the slope as follows:

$$\text{slope} = \frac{mKG_0m}{1 + KG_0m}. \tag{2.10}$$

Therefore, if m is known, K can be calculated. In CD systems, inclusion complexes show the same stoichiometry (CD:G molar ratios $= 1{:}1$, $m = 1$). Generally, the strength or magnitude of guest molecule binding within the CD cavity can be described by K_S, which can be calculated from the initial straight-line portion in the phase solubility diagram [15] using the following equation:

$$K_{1:1} = K_S = \frac{\text{slope}}{G_0(1 - \text{slope})}. \tag{2.11}$$

Following an exhaustive examination involving different CDs and several different drugs, Carrier *et al.* [16] suggested some guidelines for improving the bioavailability by increasing the solubility. They stated the following considerations for obtaining the best outcome: the utilization of preformed complexes rather than physical mixtures, guest hydrophobicity (log $P > 2.5$), low guest solubility (typically <1 mg mL^{-1}), moderate binding constant (<5000 M^{-1}), low dose (<100 mg), and low CD:drug ratio (2:1). The stability values for several CD inclusion guests are compiled in Table 2.1. Generally speaking, a K_S from ~100 to 5000 M^{-1} appears to be suitable for practical applications. The values are characteristic of weak intermolecular interactions. It is remarkable that there is no obvious correlation between the chemical properties (functional groups) of the guest molecules and the stability constants. Inclusion therefore does not seem to depend primarily on the character of the guest. Very labile complexes result in the premature release of the guest (due to weak interactions, resulting in an insignificant improvement in solubility), and very stable complexes result in a retarded or incomplete release of the guest, and consequently absorption is hindered. However, in some cases, even with small K_S values, complexation confers better physicochemical, pharmaco-technical, and biopharmaceutical properties on drugs and/or other molecules.

2.4 Methods for inclusion complex preparation

2.4.1 *Coprecipitation*

The coprecipitation process is the simplest, fastest, and cheapest methodology; hence, it is suitable for industrial-scale applications. This method can be applied in two ways: (1) Coprecipitation based on phase solubility. In this method, the solid inclusion compound is isolated from the saturated aqueous solution. The amount of host and guest to be mixed in water can be estimated from the descending curvature of a B_S-type phase solubility diagram, where there is no remaining undissolved guest and the CD is still within its solubility limit. The components are added to water and shaken until the solubility equilibrium is reached or dissolved in hot water and cooled slowly. The inclusion compound precipitates as a microcrystalline powder and is separated by filtration and dried. This method is not applicable to a system with an A-type phase solubility diagram because of the formation of a soluble inclusion compound. It is also unsuitable for large-scale preparations because large quantities of water must be removed, which is time consuming and expensive. (2) Coprecipitation not based on phase solubility. In this method, the guest is dissolved in organic solvents such as diethyl ether, chloroform, or benzene, and the appropriate amount of CD is dissolved in water and added with agitation. When crystallization occurs on cooling, crystals are washed with another organic solvent to remove nonincluded guests, and crystals are dried. This method is useful for substances that are not water soluble, but it results in a poor yield since the organic solvent used as the precipitant competitively inhibits the inclusion.

2.4.2 *Freeze-drying/lyophilization*

In this process, the guest molecule is dissolved in water (using ammonia if necessary to dissolve poorly soluble, weakly acidic compounds in aqueous solution), and the CD is then added at the correct proportion and dissolved by stirring. The mixture is freeze-dried and washed with diethyl ether, and the residue dried under vacuum. In certain circumstances, the last two steps can be omitted. This method is more suitable for water-soluble or thermosensitive guests, since CDs and guests should be dissolved in water and frozen before drying. Complexes prepared in this manner often have structures with channels [17].

2.4.3 *Spray-drying*

Spray-drying is most commonly used for the preparation of microparticle powders to be used as dry powder inhalers. The desired amount of guest molecule and CD are dissolved in water. Small quantities of organic solvent may be used to obtain limpid solutions before drying. Once solutions are transparent, they are atomized into a drying chamber using a spray nozzle. The spray-dryer is operated under the most appropriate conditions (e.g., inlet temperature, sample feeding speed). As reported by Cabral *et al.* [18], a beclomethasone dipropionate: γ-cyclodextrin (BDP:γ-CYD) complex was obtained by spray-drying in a study of particle-size distribution, shape and morphology, moisture content, uniformity in the BDP content of formulations, and spray-drying conditions (solution flow speed, T_{in} and T_{out}) with suitable characteristics for lung delivery. For thermosensitive guests, inclusion complexes should be made and then spray-dried using a typical spray-drying machine or spray-freeze-drying machine.

For samples made of small particles that cannot be dissolved, high-pressure homogenization can be useful during sample pretreatment because it produces high-quality drug nanoparticles on an industrial scale, and this nanonization technique can overcome limitations for guest solubilization [19].

2.4.4 *Kneading*

Guest compound in a liquid or solution state is added to a slurry of CD and kneaded thoroughly (in a mortar) to obtain a paste which is then dried. The solid is washed with a small amount of solvent to remove the adsorbed free guest component from the inclusion compound and dried under vacuum. This method is particularly useful for poorly water-soluble guests, since the guest dissolves slowly with the formation of inclusion compounds, and inclusion formation occurs at high yield. However, with some molecules, reproducibility is poor, and kneading is unsuitable for large-scale preparation [20].

2.4.5 *Cogrinding*

Cogrinding is a method for preparing samples in the solid state by grinding them together using a vibrating rod mill. This method is simple and rapid, and samples are mixed when still in powder form; hence, stability in solution need not be considered. The particle size of solid drugs is reduced by grinding, and the dissolution rate and crystallinity of the active ingredients (APIs) is altered to fit into the phase transitions of polymorphs and the molecular interactions in the solid state. The efficiency of grinding depends on the intensity of mechanical stress and the additives applied. For example, the cogrinding of ibuprofen with β-CD led to a positive change in its molecular state and transformed crystals into an amorphous state, resulting in improved solubility and a higher rate of dissolution for the drug. The stability of the amorphous complex has been investigated as a factor of relative humidity and temperature [21].

The effectiveness of mechanical grinding was investigated for complexation between gemfibrozil and dimethyl-β-CD by Aigner *et al.* [22]. Coground mixtures of gemfibrozil and dimethyl-β-CD at a

1:1 molar ratio were prepared in an agate mortar, and examination showed that gemfibrozil was complexed step by step during this procedure. The process could be described by a linear equation for up to 30 min of cogrinding, after which the proportion of complexed gemfibrozil did not change upon further increasing the grinding time. Based on these investigations, a solvent-free cogrinding method has been suggested for production of the complex.

2.5 Analysis of inclusion complexes

Assessment of the formation of a CD inclusion complex and its full characterization is not a simple task and often requires the use of different analytical methods, the results of which must be combined and examined together since each method explores a particular feature of the inclusion complex. The concomitant use of different techniques can allow a better and more in-depth understanding of host–guest interactions, and helps in the selection of the most appropriate CD for a given guest molecule. The different methods available are generally based on the detection of variation in any suitable physical or chemical property of the guest as a consequence of inclusion complex formation. Obviously, it is essential that the observed variation is large enough to be detected or estimated with sufficient precision. Moreover, any measurement method suffers from its own drawbacks, which should be well understood and held in consideration, in order to evaluate their effects on the reliability of the results.

2.5.1 *Thermoanalytical techniques*

2.5.1.1 *Differential scanning calorimetry (DSC)*

DSC is a very powerful analytical tool for characterization of substances in the solid state due to its ability to provide detailed information about both physical and energetic properties. Comparison of thermal curves of single components, their physical mixtures, and the presumed inclusion compounds can provide insight into the solid-state modifications and interactions between the components resulting from procedures used for complex preparation and/or of the actual inclusion complex formation. For example, Pralhad and Rajendrakumar applied DSC to evaluate an inclusion complex of quercetin–CD and assessed the liberation of crystal water, general water loss, and drug endothermal peaks before and after inclusion to probe interactions between the components in binary/CD systems [23].

2.5.1.2 *Thermal gravimetric analysis (TGA)*

TGA allows the determination of changes in the sample weight with respect to temperature changes. Comparison of TG curves from pure guest components, their physical mixtures, and complexes can reveal changes in the weight loss profiles of the putative complex and indicate interactions between components and/or the formation of a true inclusion complex. For example, TGA was used to evaluate the stoichiometric ratio (SR, guest to host) of the guest-α-cyclodextrin (Guest-α-CD) inclusion complexes 4-cresol-α-CD, benzyl alcohol-α-CD, ferrocene-α-CD, and decanoic acid-α-CD, and SR data were in good agreement with X-ray diffraction (XRD) and NMR confirmatory experiments [24].

2.5.1.3 *Hot-stage microscopy (HSM)*

HSM is an analytical technique useful for performing solid-state physical characterization of materials as a function of temperature. HSM is often used as a complementary thermal technique to DSC analysis, to corroborate the results and/or to help explain the nature of the thermal effects observed in the DSC curves. For example, examination by HSM of physical mixtures and kneaded systems of griseofulvin with hydroxypropyl-β-CD revealed that hydroxypropyl-β-CD was dissolved in the droplets of melted griseofulvin, but did not show any interaction between melted griseofulvin and CD particles. This confirmed the results of DSC analysis, in which the presence of a drug-melting peak in the thermal curves of all binary systems suggested the absence of any griseofulvin–CD inclusion compound in the solid state [25].

2.5.2 *X-ray diffraction*

X-ray diffraction is a technique commonly used for identification and characterization of crystalline materials.

2.5.2.1 *Powder X-ray diffraction (PXRD)*

Unlike the single crystal X-ray diffraction technique, PXRD can be performed on finely ground and homogenized samples. It is widely used to rapidly identify unknown crystalline substances, and to determine the degree of crystallinity or amorphization of the examined samples. PXRD has the advantage of not requiring any pretreatment of the sample, and that, unlike thermal methods of analysis, the sample does not suffer any chemical-physical changes during data collection; hence, it can be reutilized for other analyses. When putative complexes are obtained as an amorphous powder, which frequently happens when using cogrinding, freeze-drying, or spray-drying preparation methods, PXRD is not suitable for providing structural information, since it is not able to discriminate whether the obtained product is a true inclusion complex or a homogeneous dispersed mixture of the two amorphous components. However, the technique can be useful for obtaining insight about the intensity of guest molecule–CD interactions, based on the degree of amorphization in their interacting mixtures. A careful analysis of PXRD patterns of pure ketoprofen and methyl-CD, and their physical mixture and complex prepared using supercritical carbon dioxide, showed that while the physical mixture pattern was the simple superimposition of those of the pure components, the complex pattern resembled that of the amorphous CD. This indicated the disappearance of the crystalline drug in the sample, and even though this could not distinguish between drug amorphization and inclusion into the CD cavity, the partial recrystallization of the CD during the supercritical processing supported the formation of the CD complex [26].

The ability to exploit the PXRD technique for crystal structure determination of CD complexes has been demonstrated in the case of the *p*-aminobenzoic acid–CD inclusion complex, whose structure was solved directly from PXRD data using the direct space approach and refined with Rietveld refinement techniques [27]. Crystallographic data obtained by PXRD were in agreement with those obtained by single crystal X-ray diffraction and indicated that the complex crystallized in the monoclinic P21 space group, with the amino group of the guest located at the wide side of the β-CD cavity, forming hydrogen bonds with β-CD, and the carboxyl group located at the narrow side [27].

2.5.2.2 *Single-crystal X-ray diffraction (SCXRD)*

Single-crystal X-ray diffraction (SCXRD) is a nondestructive analytical technique which provides detailed data allowing crystal structure determination of samples, including unit cell dimensions, the position of all atoms within the crystal lattice, bond lengths, and bond angles. Using this technique, a three-dimensional structure of the whole molecule can be obtained, which provides information on molecular identity, conformation, and packing. However, this technique requires single stable crystals of suitable dimensions, generally between 80 and 250 μm in size, and therefore its applicability to the study of cyclodextrin complexes is rather limited. In fact, only a small number of cyclodextrin complexes have been obtained as crystalline compounds forming single crystals of suitable dimensions to be subjected to a single-crystal X-ray structure analysis.

In a recent study of inclusion complexes of essential oils with CDs, the crystal structures of both borneol enantiomers present in α-CD and β-CD were determined by SCXRD. Single crystals were prepared by gradually cooling to room temperature, over 7 days, equimolar aqueous solutions previously heated to 70°C and stirred for 4 h. Analysis of the collected data indicated that both CDs formed head-to-head dimers arranged in a "chess-board" crystal packing mode; in the case of inclusion complexes with α-CD, a guest molecule is accommodated inside the dimeric cavity, while in the case of the inclusion complexes with β-CD, a guest molecule is located inside the dimeric cavity and two additional guest molecules are located at the rims of the primary hydroxyls of the dimer, forming H-bonds with both water molecules and hydroxyl groups of the β-CDs of adjacent dimers. Moreover, no significant differences in the inclusion geometry and crystal packing were observed between inclusion complexes of borneol enantiomers with the same host CD, indicating, in this case, the inability of CDs to undergo enantiodifferentiation [28].

2.5.3 *Spectroscopic techniques*

All spectroscopic methods are based on the measurement of variation, upon inclusion complex formation, in a given property (absorbance, NMR shift, fluorescence intensity, etc.) of the system. Therefore, it is necessary that this variation can be detectable with sufficient precision. Moreover, these methods usually work at a fixed guest concentration while varying the concentration of the CD host, thus requiring the preparation of multiple sample solutions. This is time consuming and can use lots of and material, which may be a disadvantage during the initial development phases.

2.5.3.1 *Fourier transform infrared* (*FTIR*) *spectroscopy*

Fourier transform infrared (FTIR) spectroscopy is useful to identify which vibrational modes of the drug and the CD are disturbed during the inclusion process, thereby indicating interactions between the molecules in the solid state. Changes in the characteristic bands of the guest molecule, such as disappearance, broadening, variation in peak intensity and/or a shift in wave number, can be indicative of complex formation. Such changes could be the result of restriction of the stretching vibrations of the guest molecule caused by inclusion in the CD cavity and/or weakening of the interatomic bonds as a consequence of an altered environment around these bonds upon complexation. The wide diffusion, limited cost, and high sensitivity and selectivity of this analytical technique are among the main advantages of this technique, as is a relatively fast data acquisition time. However, a possible disadvantage is the sample preparation required for analysis. The classic FTIR technique involves dispersing the sample in KBr powder and pressing the mixture to form a transparent pellet. In addition to the difficulties associated with making suitable pellets, variation in mixing and in the dimensions of the resulting pellets makes this technique less useful for quantitative measurements. Physical–chemical modification of the sample can occur during the preparation of KBr disks, and their high hygroscopicity has to be taken into account.

2.5.3.2 *Attenuated total reflectance (ATR)-FTIR spectroscopy*

Attenuated total reflectance (ATR)-FTIR spectroscopy has certain important advantages over the traditional FTIR technique, since no sample preparation is required, and spectra are obtained without dispersion of the sample in KBr disks. Moreover, the lack of sample manipulation assures greater rapidity in the measurement process, and higher reproducibility of the spectra, making ATR-FTIR suitable for identifying differences in solid-state forms, including hydrated and polymorphic forms, and for the identification and characterization of pharmaceutical formulations [29].

The ATR-FTIR technique has been used to investigate complex formation in solid-state samples of genistein with nonionic amphiphilic CDs and to characterize the functional groups responsible for complex stability; monitoring of the most significant changes (broadening, shift, variation in relative intensity) in spectral bands corresponding to drug functional groups observed when passing from a physical mixture to the complex indicated changes, upon complexation, in bond strength and length, due to specific host–guest interactions. In particular, ATR-FTIR studies indicated the lack of inclusion of guest molecules inside the host macrocycle and suggested hydrogen-bonding interactions between the drug and the hydrophilic PEG chains of the macrocycle [30].

The formation of a resveratrol–sulfobuthylether-β-CD complex has been investigated by ATR-FTIR spectroscopy, and the results indicated the roles of different functional groups of guest and host molecules in the inclusion process based on significant changes in the vibrational features of the complex with respect to those of the individual components and the physical mixture; in particular, the complexation mechanism was monitored through the decomposition of the OH stretching ATR-FTIR band.

2.5.3.3 *Raman spectroscopy*

In general, Raman spectroscopy, a form of molecular spectroscopy that observes inelastically scattered light, has proved to be a powerful technique for investigation and identification of encapsulation phenomena and for monitoring changes in molecular bond structure, since it can provide valuable and specific molecular information about the interaction mechanisms underlying molecular recognition processes [31]. In comparison to FTIR spectroscopy, Raman spectroscopy has some major advantages, since it measures light scattered from a sample rather than light absorbed by a sample. Consequently, Raman spectroscopy needs little or no sample preparation, and it is insensitive to water absorption bands. Regions of Raman spectra that are free of bands corresponding to CDs can be used as windows for investigating changes in relevant guest bands resulting from putative complex formation, in particular stretching vibrations of double bonds (C=O) and aromatic C–H bonds.

Raman spectroscopy has been successfully used to study inclusion complexation of curcumin with natural β-CD and γ-CD hosts. The curcumin Raman spectrum was not affected by the presence of β-CD at any amount of the host, suggesting that no dye encapsulation occurred. By contrast, Raman spectra provided specific structural information in the presence of γ-CD from ligand signals, which indicates that encapsulation took place at the level of the aromatic rings, through H-bonds, and that a tautomerization from the planar keto enol form to the nonplanar diketo form of curcumin also occurred [32].

2.5.3.4 *Circular dichroism spectroscopy*

Circular dichroism is a powerful technique for proving CD inclusion complexation of both chiral and nonchiral guest molecules and provides information on the structure of the complex in aqueous solution. In the case of chiral guest molecules, changes in their circular dichroism spectra may be detected that are attributed to increased optical activity induced by the formation of inclusion complexes with CDs. The effect is only observed when the chromophore moiety of the guest molecule is included in the CD cavity. An outer surface association of a potential guest with the CD may give rise to some variation in other spectral properties but not to induced Cotton effects. Fumes and colleagues reported an induced circular dichroism (ICD) signal intensity, with Cotton effects, which demonstrated that the electric dipole moment of the molecule coincides with the β-CD *n*fold axis, confirming axial inclusion and a perpendicular orientation between the guest molecule and the β-CD axis [33].

2.5.3.5 *Ultraviolet (UV)–visible spectroscopy*

UV–visible spectroscopy is a simple, economical, fast and useful method for studying the formation of host–guest complexes in solution. When complex formation with cyclodextrin alters the original visible or UV absorption spectrum of the guest, a bathochromic shift and/or band broadening usually occurs. The shift in the UV absorption maximum upon complex formation may be explained by partial shielding of the excitable electrons in the cyclodextrin cavity. Ateba *et al.* characterized the interaction of Mammea A/AA (MA) with β-CD by UV–visible absorption spectroscopy together with other methods. From a comparison of spectra of MA in solution alone and in the presence of an increasing amount of β-CD, a systematic blue shift and hyperchromic effect was attributed to the formation of an inclusion complex with β-CD [34].

2.5.4 *Photoacoustic spectroscopy*

Photoacoustic spectroscopy is an innovative and low-cost technique that provides many advantages over the aforementioned techniques because it is nondestructive and requires minimal sample preparation.

2.5.4.1 *Scanning electron microscopy (SEM)*

SEM is used in the solid-state characterization of guest-CD systems since it allows an in-depth investigation of morphological aspects of raw materials, their physical mixtures, and putative complexes. Although this technique is not capable of confirming true inclusion complex formation, it can provide evidence of morphological changes related to interactions between the components and reveal the existence of a single component in preparations. Before examination with a conventional electron microscope, samples are sputter coated with gold or gold–palladium alloy in vacuum to render them electrically conductive. This limits the possible applications of SEM, since specimens could be modified and compromised by such treatment. Environmental scanning electron microscopy (ESEM) represents an evolution of conventional SEM because it works under a gaseous environment in the specimen chamber, which enables the examination of practically any kind of specimen, wet or dry, in its natural state.

For example, ESEM analysis was used to investigate the possible morphological changes of spray-dried prilocaine. The HCl–triacetyl-β-CD product was compared with its pure components and their simple physical mixtures [19]; the pure drug formed crystalline needle-shaped sticks, with rather homogeneous dimensions, and pure CDs formed rhomboid-shaped crystals of different dimensions; the physical mixture was characterized by the presence of unchanged crystal particles of CDs mixed with typical small crystalline sticks of the drug; interestingly, a remarkable morphological change was observed in the spray-dried system, which revealed an amorphous homogeneous product characterized by particles of irregular shape and dimensions and with a fluffy appearance that were completely different form the mother components. This result, even though scarcely conclusive, indicated the presence of a single solid phase in the spray-dried system and consequently implied the formation of an inclusion complex, further supporting the results of DSC, PXRD, and FTIR analyses [19].

SEM images of pure sulfamethazine, β-CD, methyl-CD, and their binary systems prepared by physical mixing and freeze-drying methods revealed characteristic drug crystals and parallelogram-shaped β-CD crystals, but spherical amorphous particles of methyl-CD in guest–CD physical mixtures. Meanwhile, particles of freeze-dried products displayed significant morphological changes in their shape and aspect with respect to the pure components, indicating a simple change in the crystalline habits of the systems or the presence of a new solid phase as a consequence of complex formation [35]. The surface structures of raw barbigerone, hydroxypropyl-β-CD and their physical mixture and putative complex was examined by SEM. Both typical needle-like drug crystals of different sizes and spherical amorphous particles of hydroxypropyl-β-CD were detected in the physical mixture, while the putative complex appeared as irregular blocky particles, and the original morphology of both components disappeared completely, indicating complex formation [36]. Consistently, characterization of an inclusion complex of S(–)bvc in 2-hydroxypropyl-β-cyclodextrin (HP-β-CD) by SEM showed structural changes in the complex [37].

2.5.4.2 *Transmission electron microscopy (TEM)*

TEM is used in the characterization of CD inclusion complexes in aqueous solution. TEM instruments are capable of imaging at a significantly higher resolution than light microscopes, owing to the small de Broglie wavelength of electrons. This makes it possible to examine fine details, down to a single column of atoms, which is thousands of times smaller than the smallest object resolvable by a light microscope. TEM is used in a wide range of scientific fields in the physical, chemical, and biological sciences.

TEM images can show the size and morphology of CD complex particles. There can be discrepancies in size when measuring β-CD inclusion complexes by light diffraction and TEM. This could be due to the preparation of samples for TEM, which charges the particles on the grid in order for them to adhere to the sample, potentially causing the agglomerated particles to separate into smaller particles in the TEM images [38].

The inclusion complex of citral and monochlorotriazinyl-β-cyclodextrin was characterized by TEM [39], and the result showed that inclusion complex aggregates formed a bicontinuous "worm-type" pore system.

The surface character of γ-cyclodextrin-stabilized solid lipid nanoparticles (SLNs) prepared using a stearic acid–γ-CD inclusion complex was studied by TEM, and spherical particles were observed [40].

2.5.5 *Nuclear magnetic resonance* (*NMR*)

NMR is one of the most useful techniques for studying interactions of CDs with guest compounds. It is relatively easy to perform, experiments are fast, and it is the only technique that provides information on the correct orientation of the guest molecule inside the cavity and also on other important parameters related to the physicochemical characteristics of the inclusion complexes.

2.5.5.1 *¹H-NMR spectroscopy*

The simplest NMR experiment for generating rapid, direct evidence of the inclusion of a guest in the CD cavity is the observation of differences in the proton (^{1}H-NMR) or carbon (^{13}C-NMR) chemical shifts (δ) between the free guest and host species and the presumed complex. Analysis of chemical shift changes of both host and guest molecules not only provides evidence of complex formation but also supplies useful information about the stoichiometry, stability, mechanism, and geometry of the complex. Measurement of chemical shift changes of the guest as a function of increasing CD concentration (NMR titration) allows the evaluation of the complex association constant and provides insight into the stoichiometry and conformation of the formed complex. The main drawback of this technique is the poor solubility of the samples in deuterated water (^{1}H-NMR) or in water (^{13}C-NMR), often requiring the use of other solvents which could modify the host–guest interactions with respect to simple aqueous medium. Moreover, the induced shifts are sometimes too small and may suffer from signal broadening. If a guest molecule is incorporated into the cavity, the screening constants of the cyclodextrin protons inside the cavity (H3 and H5) should be sensitive to the changed environment, but those of the outer protons (H1, H2, and H4) are not, resulting in chemical shift changes or the inside protons [41].

2.5.5.2 ^{13} C-NMR spectroscopy

From ^{13}C-NMR spectra (which may be obtained in aqueous solution), atoms of the guest molecule involved in stabilizing the complex can be identified, as can their orientation. The more intense the interaction between the guest molecule and the wall of the CD cavity, the higher is the shift of the carbon atoms involved [42]. According to a report from Darekar and colleagues, β-CD ^{13}C-NMR signals at δ 102.42, δ 82.03, δ 73.52, δ 72.89, δ 72.51, and δ 60.41 correspond to C-1, C-4, C-2, C-5, C-3, and C-6, and comparison of the ^{13}C-NMR signals of the complex with those of the isolated drug indicated that all drug carbon atoms underwent an upward chemical shift ($\Delta\delta$) [43].

2.6 Conclusion

Although the mechanisms and forces that act to achieve complexation remain the subject of debate and controversy, CDs and their derivatives have been successfully employed to prepare complexes for use in biomedicines, foods, textiles, and cosmetotextiles. When selecting a preparation method, the inclusion efficiency, preparation process complexity, the final application of the inclusion complex should be considered together. Finally, a combination of analytical methods should be adopted to evaluate putative inclusion complexes.

References

1. Lewis EA, Hansen LD. (1973). Thermodynamics of binding of guest molecules to α- and β-cyclodextrins. *Journal of the Chemical Society, Perkin Transactions*, 2(15), pp. 2081–2085.
2. Gelb RI, Schwartz LM, Johnson RF, Laufer DA. (1979). The complexation chemistry of cyclohexaamyloses. 4. Reactions of cyclohexaamylose with formic, acetic, and benzoic acids and their conjugate bases. *Journal of the American Chemical Society*, 101(7), pp. 1869–1874.
3. Hersey A, Robinson BH, Kelly HC. (1986). Mechanism of inclusion-compound formation for binding of organic dyes, ions and surfactants to α-cyclodextrin studied by kinetic methods based on competition experiments. *Journal of the Chemical Society, Faraday Transactions 1: Physical Chemistry in Condensed Phases*, 82(5), pp. 1271–1287.

4. Komiyama M, Bender ML. (1978). Importance of apolar binding in complex formation of cyclodextrins with adamantanecarboxylate. *Journal of the American Chemical Society*, 100(7), pp. 2259–2260.
5. Shehatta I. (1996). Thermodynamics of macrocyclic compounds I. Inclusion complexes of α- and β-cyclodextrins with some nonelectrolytes in water. *Reactive and Functional Polymers*, 28(2), pp. 183–190.
6. Ueda H, Perrin JH. (1986). A microcalorimetric investigation of the binding of flurbiprofen to cyclodextrins. *Journal of Pharmaceutical and Biomedical Analysis*, 4(1), pp. 107–110.
7. Rekharsky MV, Nemykina EV, Eliseev AV, Yatsimirsky AK. (1992). Thermodynamics of molecular recognition of nitrogen heterocycles: Part 1. Interaction of imidazole and imidazolium cation with α-cyclodextrin and β-cyclodextrin. *Thermochimica Acta*, 202, pp. 25–33.
8. Guo QX, Zheng XQ, Luo SH, Liu YC. (1996). Enthalpy-entropy compensation in the inclusion of 1-substituted naphthalenes by beta-cyclodextrin in water. *Chinese Chemical Letters*, 7(4), pp. 357–360.
9. Blyshak LA, Patonay WG. (1990). Evidence for non-inclusional association between alpha-cyclodextrin and polynuclear aromatic-hydrocarbons. *Analytica Chimica Acta*, 239(2), pp. 239–243.
10. Harata K, Tsuda K, Uekama K, Otagir M, Hirayama F. (1988). Complex formation of hexakis(2,3,6-tri-O-methyl)-α-cyclodextrin with substituted benzenes in aqueous solution. *Journal of Inclusion Phenomena*, 6(2), pp. 135–142.
11. Liu Y, Zhang YM, Sun SX, Zhang ZH, Chen RT. (1997). Molecular recognition study on supramolecular system. 5. Molecular recognition thermodynamics of amino acids by mono-[6-(1-pyridinio)-6-deoxy]-alpha- and gamma-cyclodextrins. *Acta Chimica Sinica*, 55(8), pp. 779–785.
12. Bonomo RP, Cucinotta V, D'Alessandro F, Impellizzeri G, Maccarrone G, Vecchio G, Rizzarelli E. (1991). Conformational features and coordination properties of functionalized cyclodextrins. Formation, stability, and structure of proton and copper(II) complexes of histamine-bearing beta-cyclodextrin in aqueous solution. *Inorganic Chemistry*, 30(13), pp. 2708–2713.
13. Lammers JNJJ, Koole JL, Hurkmans J. (1971). Properties of cyclodextrins. Part VI. Water-soluble cyclodextrin-derivatives. Preparation and analysis. Starch — Stärke, 23(5), pp. 167–171.
14. Bertrand GL, Faulkner JR, Han SM, Armstrong DW. (1989). Substituent effects on the binding of phenols to cyclodextrins in aqueous solution. *The Journal of Physical Chemistry*, 93(18), pp. 6863–6867.
15. Higuchi T, Connors KA. (1965). Phase solubility techniques. *Advances in Analytical Chemistry and Instrumentation*, 4, pp. 117–212.
16. Carrier RL, Miller LA, Ahmed I. (2007). The utility of cyclodextrins for enhancing oral bioavailability. *Journal of Controlled Release*, 123(2), pp. 78–99.
17. Inoue Y, Watanabe S, Suzuki R, Murata I, Kanamoto I. (2015). Evaluation of actarit/γ-cyclodextrin complex prepared by different methods. *Journal of Inclusion Phenomena and Macrocyclic Chemistry*, 81(1), pp. 161–168.

18. Cabral-Marques H, Almeida R. (2009). Optimisation of spray-drying process variables for dry powder inhalation (DPI) formulations of corticosteroid/cyclodextrin inclusion complexes. *European Journal of Pharmaceutics and Biopharmaceutics*, 73(1), pp. 121–129.
19. Chen H, Chalermchai K, Yang X, Chang X, Gao J. (2011). Nanonization strategies for poorly water-soluble drugs. *Drug Discovery Today*, 16(7–8), pp. 354–360.
20. Sá Couto A, Salústio P, Cabral-Marques H. (2014). Cyclodextrins. In: Ramawat GK, Mérillon JM (Eds.), *Polysaccharides: Bioactivity and Biotechnology*. Cham: Springer International Publishing, pp. 1–36.
21. Wang Q, Li S, Che X, Fan X, Li C. (2010). Dissolution improvement and stabilization of ibuprofen by co-grinding in a β-cyclodextrin ground complex. *Asian Journal of Pharmaceutical Sciences*, 5(5), pp. 188–193.
22. Aigner Z, Berkesi O, Farkas G, Szabó-Révész P. (2012). DSC, X-ray and FTIR studies of a gemfibrozil/dimethyl-β-cyclodextrin inclusion complex produced by co-grinding. *Journal of Pharmaceutical and Biomedical Analysis*, 57, pp. 62–67.
23. Pralhad T, Rajendrakumar K. (2004). Study of freeze-dried quercetin–cyclodextrin binary systems by DSC, FT-IR, X-ray diffraction and SEM analysis. *Journal of Pharmaceutical and Biomedical Analysis*, 34(2), pp. 333–339.
24. Bai Y, Wang J, Bashari M, Hu X, Feng T, Xu X, Jin Z, Tian Y. (2012). A thermogravimetric analysis (TGA) method developed for estimating the stoichiometric ratio of solid-state α-cyclodextrin-based inclusion complexes. *Thermochimica Acta*, 541, pp. 62–69.
25. Mura P. (2015). Analytical techniques for characterization of cyclodextrin complexes in the solid state: a review. *Journal of Pharmaceutical and Biomedical Analysis*, 113, pp. 226–238.
26. Banchero M, Ronchetti S, Manna L. (2013). Characterization of ketoprofen/methyl-β-cyclodextrin complexes prepared using supercritical carbon dioxide. *Journal of Chemistry*, pp. 1–8.
27. Guo P, Su Y, Cheng Q, Pan Q, Li H. (2011). Crystal structure determination of the β-cyclodextrin-p-aminobenzoic acid inclusion complex from powder X-ray diffraction data. *Carbohydrate Research*, 346, pp. 986–990.
28. Christoforides E, Mentzafos D, Bethanis K. (2015). Structural studies of the inclusion complexes of the (+)- and (−)-borneol enantiomers in α- and β-cyclodextrin. *Journal of Inclusion Phenomena and Macrocyclic Chemistry*, 81(1), pp. 193–203.
29. Elbashir AA, Dsugi NF, Mohmed TO, Aboul-Enein HY. (2014). Spectrofluorometric analytical applications of cyclodextrins. *Luminescence: The Journal of Biological and Chemical Luminescence*, 29(1), pp. 1–7.
30. Cannavà C, Crupi V, Ficarra P, Guardo M, Majolino D, Mazzaglia A, Stancanelli R, Venuti V, (2010). Physico-chemical characterization of an amphiphilic cyclodextrin/genistein complex. *Journal of Pharmaceutical and Biomedical Analysis*, 51(5), pp. 1064–1068.

31. Edward HW, Stinne WH, Martin C, Thomas SH, Sune DN, Jan OJ, Eric WW, Lasse J, Amar HF. (2009). Determination of binding strengths of a host–guest complex using resonance Raman scattering. *Journal of Physical Chemistry A*, 113(34), pp. 9450–9457.

32. López-Tobar E, Blanch GP, Ruiz del Castillo ML, Sanchez-Cortes S. (2012). Encapsulation and isomerization of curcumin with cyclodextrins characterized by electronic and vibrational spectroscopy. *Vibrational Spectroscopy*, 62, pp. 292–298.

33. Fumes BH, Guzzo MR, Machado AEH, Okano LT. (2016). Study of the mode of inclusion for 7-hydroxyflavone in β-cyclodextrin complexes. *Journal of the Brazilian Chemical Society*, 27(2), pp. 382–391.

34. Ateba BA, Lissouck D, Azébazé A, Ebelle CT, Nassi A, Ngameni E, Duportail G, Mbazé L, Kenfack CA. (2016). Characterization of Mammea A/AA in solution and in interaction with β-cyclodextrin: UV–visible spectroscopy, cyclic voltammetry and DFT-TDDFT/MD study. *Journal of Molecular Liquids*, 213, pp. 294–303.

35. Zoppi A, Delrivo A, Aiassa V, Longhi MR. (2013). Binding of sulfamethazine to β-cyclodextrin and methyl-β-cyclodextrin. *AAPS PharmSciTech*, 14(2), pp. 727–735.

36. Qiu N, Cheng X, Wang G, Wang W, Wen J, Zhang Y, Song H, Ma L, Wei Y, Peng A, Chen L. (2014). Inclusion complex of barbigerone with hydroxypropyl-β-cyclodextrin: preparation and *in vitro* evaluation. *Carbohydrate Polymers*, 101, pp. 623–630.

37. Moraes CM, Araújo DRD, Issa MG, Ferraz HG, Yokaichiya F, Franco MK, Fraceto LF. (2009). Inclusion complex of S(−) bupivacaine and 2-hydroxypropyl-β-cyclodextrin: study of morphology and cytotoxicity. *Revista de Ciências Farmacêuticas Básica e Aplicada*, 27(3), pp. 207–212.

38. Hill LE, Gomes C, Taylor TM. (2013). Characterization of beta-cyclodextrin inclusion complexes containing essential oils (trans-cinnamaldehyde, eugenol, cinnamon bark, and clove bud extracts) for antimicrobial delivery applications. *LWT — Food Science and Technology*, 51(1), pp. 86–93.

39. Zhu G, Feng N, Xiao Z, Zhou R, Niu Y. (2015). Production and pyrolysis characteristics of citral-monochlorotriazinyl-β-cyclodextrin inclusion complex. *Journal of Thermal Analysis and Calorimetry*, 120(3), pp. 1811–1817.

40. Negi JS, Chattopadhyay P, Sharma AK, Ram V. (2014). Preparation of gamma cyclodextrin stabilized solid lipid nanoparticles (SLNS) using stearic acid–γ-cyclodextrin inclusion complex. *Journal of Inclusion Phenomena and Macrocyclic Chemistry*, 80(3), pp. 359–368.

41. Marcolino VA, Zanin GM, Durrant LR, de Toledo Benassi M, Matioli G. (2011). Interaction of curcumin and bixin with beta-cyclodextrin: complexation methods, stability, and applications in food. *Journal of Agricultural and Food Chemistry*, 59(7), pp. 3348–3357.

42. Schneider H-J, Hacket F, Rüdiger V, Ikeda H. (1998). NMR studies of cyclodextrins and cyclodextrin complexes. *Chemical Reviews*, 98(5), pp. 1755–1786.
43. Darekar T, Aithal KS, Shirodkar R, Kumar L, Attari Z, Lewis S. (2016). Characterization and *in vivo* evaluation of lacidipine inclusion complexes with β-cyclodextrin and its derivatives. *Journal of Inclusion Phenomena and Macrocyclic Chemistry*, 84(3), pp. 225–235.

3. Applications in Food

Chao Yuan*, Wangyang Shen†, Bo Yu‡,
and Xing Zhou§

*School of Food Science and Engineering,
Qilu University of Technology,
Jinan 250353, China

†School of Food Science and Engineering,
Wuhan Polytechnic University,
Wuhan 430000, China

‡College of Chemical Engineering and Food Science,
Hubei University of Arts and Science,
Xiangyang 441053, China

§School of Food Science and Technology,
Jiangnan University, Wuxi 214122, China

3.1 Introduction

Cyclodextrins (CDs) are able to form host–guest inclusion complexes with a wide variety of hydrophobic guest molecules. One or two guest molecules can be encapsulated by one, two, or three CDs. The physicochemical and biological properties of the guest molecule can be changed after complex formation to give beneficial properties. For example, formation of inclusion complexes can enhance the solubility and/or stability of guest molecules and may also allow control of their release [1]. The most common and commercially available natural CDs are α-, β-, and γ-CDs, which are composed of six, seven, and eight glucose units, respectively, although the food industry almost exclusively uses β-CD due to its low price and easy accessibility.

Natural CDs, in particular β-CD, are of limited aqueous solubility; hence, they may also be poorly soluble, and solid CD complexes can be precipitated from water and other aqueous systems. Substitution of any of the hydrogen bond-forming hydroxyl groups located on the edge of the CD cone, even by lipophilic functional groups, results in significant improvement in their aqueous solubility. CD derivatives of food industry of interest include the hydroxypropyl derivatives of β-CD and γ-CD (i.e., HP-β-CD and HP-γ-CD), the randomly methylated β-CD (RM-β-CD), sulfobutylether β-CD (SBE-β-CD), and the so-called branched CDs such as maltosyl-β-CD (G$_2$-β-CD).

CDs are nontoxic, edible, chemically stable, easy to separate, and readily available. α-CD, β-CD, and γ-CD were introduced into the generally regarded as safe (GRAS) list of the Food and Drug Administration (FDA) for use as a food additive in 2004, 2001, and 2000, respectively, and HP-β-CD is cited in the FDA's list of Inactive Pharmaceutical Ingredients. SBE-β-CD is also available in various dosages and is also listed in the FDA's compilation of Inactive Pharmaceutical Ingredients [2].

CDs are useful functional additives that are used widely in foods as wall materials for molecular encapsulation of flavors and other organic ingredients, because the central CD cavity provides a lipophilic microenvironment into which suitably sized hydrophobic food molecules may enter and reside [3, 4]. Moreover, besides their complex forming ability, they can improve the water retention and homogeneity of their components.

3.2 Application of CD complexes in foods

The application of CD-assisted molecular encapsulation in foods offers the following advantages [5]:

(1) Protection of active ingredients against oxidation, light-induced reactions, heat-promoted decomposition, loss by volatility, and sublimation.
(2) Elimination (or reduction) of undesired tastes/odors, microbiological contamination, fibers/other undesired components, and hygroscopicity.

(3) Technological advantages that include typically stable, standard-izable composition, simple dosing and handling of dry powders, reduced packing and storage costs, advanced processes, and a decrease in manpower [6].

α-CD has been allocated an Acceptable Daily Intake of "not specified" [7] based on the known current uses of α-CD prepared under good manufacturing practices as a carrier and stabilizer for flavors, colors, and sweeteners; water solubilizer for fatty acids and certain vitamins; flavor modifier in soya milk; and absorbent in confectionery products [8].

3.2.1 *CDs as flavor carriers*

Flavors are of much importance in determining consumer satisfaction, and they affect the consumption of foods [9]. During production and storage processes, flavors are often reduced by oxygen and evaporation, since most natural and artificial flavors are volatile oils or liquids [10]. To inhibit the loss or degradation of flavors during storage, it is beneficial to encapsulate volatile ingredients prior to the processing of food products. Encapsulation techniques include spray-drying, freeze-drying, cocrystallisation, and forming inclusion complexes with CDs. The formation of CD–flavor inclusion complexes offers great potential for retaining flavor compounds in multicomponent food systems [11].

Many studies on the ability of CDs to protect flavors against heat, oxygen, and evaporation have been conducted [11, 12]. For example, β-CD acts as a stabilizing or thickening agent to retain some aroma compounds in food products [40]. As a molecular encapsulant, it also improves the flavor quality and extends the preservation period compared with other encapsulants [12]. The β-CD-based encapsulation of flavors is a complexation process on a molecular scale. This process inhibits molecular interactions between the different components of natural or synthetic flavors and essential oils, encapsulating flavors into the β-CD-based inclusion complexes without changing the overall food composition [6]. The flavor load of these complexes varies from 6 to 15% (Table 3.1).

Table 3.1.   Flavor load of CDs–flavors complexes.

Name of CD–flavor inclusion complex	Flavor load in the CD-based inclusion complex (%)	Refs.
HPβ-CD–diallyl sulfide	13.9	[37]
β-CD–garlic oil	9.4	[36]
β-CD–citral	9.5	[36]
β-CD–citronellal	9.0	[36]
β-CD-β–Ionone	13.3	[36]
β-CD–linalool	10.1	[36]
β-CD–bergamott oil	9.8	[36]
β-CD–jasmin oil	11.0	[36]
β-CD–sage oil	10.5	[36]
β-CD–cinnamon oil	10.4	[36]
β-CD–orange oil	8.9	[36]
β-CD–lemon oil	10.2	[36]
β-CD–lime oil	9.6	[36]
β-CD–onion oil	10.0	[36]
β-CD–mustard oil	10.2	[36]
β-CD–marjoram oil	9.9	[36]
β-CD–basil oil	10.7	[29, 38]
β-CD–laurel leaf oil	10.8	[29, 38]
β-CD–benzaldehyde	8.7	[29, 38]
β-CD–caraway oil	10.5	[29, 38]
β-CD–carrot oil	8.8	[29, 38]
β-CD–celery oil	10.0	[29, 38]
β-CD–dill oil	6.9	[29, 38]
β-CD–sage oil	8.2	[29, 38]
β-CD–thyme oil	9.6	[29, 38]

Natural spices used for seasoning and their equivalent CD–flavor complexes have also been reviewed in previous studies (Table 3.2) [11, 13], and the results indicate that the protective effects of CD-based inclusion complexes are better than those of natural spices.

CDs are used in food formulations for flavor protection and flavor delivery [12]. They form inclusion complexes with a variety of molecules including fats, flavors, and colors. Most natural and artificial flavors are volatile oils or liquids, and complexation with CDs provides a promising alternative to the conventional encapsulation

Table 3.2. Equivalent amounts of β-CD–flavor inclusion complex and corresponding natural spices determined by sensory trials.

Foods	Equivalent of 1 g of β-CD–flavor inclusion complex (g)	Refs.
Onion	130–500	[29, 39]
Dill	150–300	[29, 39]
Garlic	33–100	[29, 39]
Cumin	3–100	[29, 39]
Marjoram	3–5	[29, 39]

technologies used for flavor protection. CDs are also used as process aids, for example, to remove cholesterol from products such as milk, butter, and eggs. CDs were reported to have a texture-improving effect on pastry and meat products. Other applications arise from their ability to reduce bitterness, ill smell, and taste and to stabilize flavors when subjected to long-term storage. Mulsions like mayonnaise, margarine, or butter creams can be stabilized with β-CD. Using β-CD, cholesterol may be removed from milk to produce dairy products low in cholesterol [14, 15].

CDs act as molecular encapsulants, protecting the flavor throughout many rigorous food-processing methods including freezing, thawing, and microwaving. As a molecular encapsulant, CDs preserve the flavor quality and quantity to a greater extent and for a longer period compared with other encapsulants, and they provide longevity to food items [16]. In Japan, CDs have been approved as "modified starch" for food applications for more than two decades, serving to mask odors in fresh foods and to stabilize fish oils. One or two European countries including Hungary have approved CDs for use in certain applications because of their low toxicity.

The complexation of CDs with sweetening agents such as aspartame stabilizes and improves the taste. It also eliminates the bitter aftertaste of other sweeteners such as stevioside, glycyrrhizin, and rubusoside. Indeed, CD itself is a promising new sweetener. Enhancement of flavor by CDs has been also claimed for alcoholic beverages such as whisky and beer [17]. The bitterness of citrus fruit juices is a major problem in the industry that is caused by the presence of limonoids (mainly limonin) and flavonoids (mainly naringin). Cross-linked CD–polymer inclusion complexes are useful for removing these bitter components.

The most prevalent use of CDs is in processes for removing cholesterol from animal products such as eggs and dairy products, and 80% of cholesterol can be removed using this approach. Free fatty acids can also be removed from fats using CDs, thus improving their frying properties (e.g., reduced smoke formation, less foaming, less browning, and less deposition of oil residues on surfaces) [15]. Fruit and vegetable juices are also treated with CDs to remove phenolic compounds that cause enzymatic browning. In juices, polyphenol oxidase converts the colorless polyphenols to colored compounds, and the addition of CDs removes this enzyme from juices by complexation. Sojo *et al.* [18] studied the effect of CDs on the oxidation of *o*-diphenol by banana polyphenol oxidase and found that CDs act as activators as well as inhibitors. By combining 1–4% CDs with chopped ginger root, Sung [19] established that it can be stored in a vacuum at cold temperature for 4 weeks or longer without browning or rotting. Flavonoids and terpenoids are beneficial for human health because of their antioxidative and antimicrobial properties, but they cannot be utilized in foodstuffs owing to their very low aqueous solubility and bitter taste. Sumiyoshi [20] discussed improving the properties of these plant components via CD complexation. CDs are used in the preparation of foodstuffs in different ways. For example, highly branched CDs are used in flour-based items such as noodles, pie doughs, pizza sheets, and rice cakes to impart elasticity and flexibility to the dough [21]. They are also used in the preparation of antimicrobial food preservatives containing *trans*-2-hexanalin in apple juice preparation [22] and in the processing of medicinal mushrooms for the preparation of crude drugs and health foods. CDs are used in the preparation of controlled-release powdered flavors and confectionery items and also in chewing gum to retain flavor for longer duration, a property highly valued by consumers [23].

From a functional standpoint, CDs can be considered as "empty capsules" of varying molecular size, and CD complexation of flavors is essentially an encapsulation process on a molecular scale.

The main unique feature of this molecular encapsulation compared with traditional encapsulation is the effective protection for all flavor constituents present in a multicomponent food system. This molecular-scale encapsulation inhibits or excludes molecular interactions between the different components of natural or synthetic composite systems such as flavor concentrates, essential oils, and oleoresins [6].

In general, dry microcrystalline CD complexes are wettable, almost odorless, nonhygroscopic powders. The crystallinity, flowing properties, and other mechanical properties of flavor inclusion complexes depend on the conditions employed in the complexation procedure (in particular on the drying processes of wet complexes).

Flavor–β-CD complexes, prepared by cocrystallization, kneading, and suspension, show remarkable resistance toward moisture sorption and clumping upon high humidity storage. The aroma and flavor load of these complexes varies in most cases between 6 and 15% w/w, and is usually is in the range of 8–10% (see Table 3.3) [24]. These

Table 3.3.   Flavor load of flavor–β-CD complexes determined by gas chromatography [25].

Name of the flavor	Flavor load of the complex (%)
Citral–CD	9.5
Citronellal	9.0
β-Ionone–CD	13.3
Linalool–CD	10.1
Linalyl-acetate–CD	9.8
L-Menthol–CD	10.6
Sage oil–CD	10.5
Cinnamon oil–CD	10.4
Jasmin oil–CD	11.0
Bergamott oil–CD	9.8
Orange oil–CD	8.9
Lemon oil–CD	10.2
Lime oil–CD	9.6
Onion oil–CD	10.0
Garlic oil–CD	9.4
Mustard oil–CD	10.2
Marjoram oil–CD	9.9

values refer to an approximately 1:1 host:guest molar ratio, since the majority of flavor components are mono- and sesquiterpenoids and phenylpropane derivatives of an average molecular weight of 120–160 (the molecular mass of β-CD is 1135).

α-CD is also a suitable agent for encapsulating flavors extracted from dried shiitake, including lenthionine, by encapsulating in a powder form via spray-drying. The retention of flavors is markedly increased by using a combination of α-CD and maltodextrin as the encapsulant [7, 26].

Binding constants of 13 different flavors (maltol, furaneol, vanillin, methyl cinnamate, cineole, citral, menthol, geraniol, camphor, nootkatone, eugenol, p-vinyl guaiacol, and limonene) to CDs (α-CD and β-CD) have been determined by UV–Vis spectrophotometry. In all cases, binding constants of flavors are larger with β-CD, due to the different sizes of the CD cavities. Additionally, there exists a relationship between the log P value of each flavor and the binding constant, proving that the driving force for host–guest complex formation is hydrophobic/hydrophilic interactions [27]. There is much interest in manipulating the complex-forming ability of CDs with a view to developing further applications [11, 28]. Recently, several reviews describing the use of CDs in food and flavor applications have been published [6, 11, 29–33]. CDs have been recommended for applications in food processing and as food additives with a variety of aims: (i) to protect lipophilic food components that are sensitive to oxygen-, light-, or heat-induced degradation; (ii) to solubilize food colorings and vitamins; (iii) to stabilize fragrances, flavors, vitamins, and essential oils against unwanted changes; (iv) to suppress unpleasant odors or tastes; and (v) to achieve a controlled release of certain food constituents.

In this regard, we cannot forget that flavor plays an important role in consumer satisfaction and influences the consumption of foods [34]. Manufacturing and storage processes, packaging materials, and ingredients in foods often influence the overall flavor by reducing aroma compound intensity or producing off-flavor components [10, 34]. To limit aroma degradation or loss during processing and storage, it is beneficial to encapsulate volatile ingredients prior to use in foods and beverages. A variety of commercial encapsulation practices are currently followed; however, those involving the formation of flavor–CD molecular inclusion complexes offer great potential for the protection of volatile and/or labile flavoring materials present in multicomponent food systems during rigorous food processing methods (cooking, pasteurization, etc.) [32, 35–37]. Similarly, CDs can eliminate some undesirable tastes, which is important because a bitter taste is the main reason for the rejection of various food products, although exceptions to this rule are rooted in many cultures. For example, in foods and beverages such as coffee, beer, and wine, a certain degree of bitterness is expected [38–42]. Bitterness, however, has proved a major limitation in the acceptance of commercial citrus juices. A commercial process is needed that removes bitter components without adding anything to the juice, while still maintaining the expected flavor and nutritional value of the product. CDs can be used for the removal or masking of undesirable components. Some foods have a peculiar smell, but when CDs are added during their manufacture, these components form CD inclusion complexes that deodorize the resultant product. For instance, this process is used for deodorizing soybean milk and soy protein and also for removing peculiar fish odors in seafood and meat products [43–45].

The formation of inclusion complexes with CDs can also protect some lipophilic food components that are sensitive to oxygen and heat- or light-induced degradation [12]. In addition, CDs protect phenolic compounds against enzymatic oxidation by forming inclusion complexes [46–49].

3.2.2 *CDs as protectants of food ingredients*

Cyclodextrin complexes are used to protect food ingredients against oxidation, light-induced decomposition, and heat-induced changes, thereby improving the shelf life of food products.

3.2.2.1 *CDs protect against oxidation*

Cyclodextrins are wall materials that form molecular capsules which entrap oxygen-sensitive food ingredients such as flavors, fatty acids, organic acids, volatile precursors, colorants, and others. The complexes formed with cyclodextrins often enhance the chemical stability of the food ingredients.

Sojo *et al.* studied the effect of cyclodextrins on the oxidation of *o*-diphenol by banana polyphenol oxidase and found that cyclodextrins act as activators as well as inhibitors [50]. By combining 1–4% CD with chopped ginger root, Sung established that it can be stored in a vacuum at cold temperatures for 4 weeks or longer without browning or rotting [19]. Flavonoids and terpenoids are beneficial for human health due to their antioxidative and antimicrobial properties, but they cannot be utilized in foodstuffs owing to their very low aqueous solubility and bitter taste. Sumiyoshi discussed improving the properties of these plant components (flavonoids and terpenoids) by cyclodextrin complexation [51].

The complete or partial entrapment of oxygen-sensitive flavors or other food ingredients such as unsaturated fatty acids and colorants by cyclodextrins can improve the chemical stability of the encapsulated agents. Such cyclodextrin-stabilized flavors were found to survive even extreme conditions, as illustrated by pure oxygen-stressed cinnamaldehyde and benzaldehyde [52]. The effect of oxygen stress on flavor identity was investigated using a Wartburg apparatus that monitors manometrically the oxygen consumption of test samples. The results showed that CD complexation almost entirely prevented these oxidizable substances from chemical alteration, even when stored in a pure oxygen atmosphere [53].

A large number of plant extracts, including coumarins, cinnamaldehydes, benzaldehydes, and flavonoids, which are employed in the food industry, also serve as major active ingredients in health-related products. However, the effectiveness of these active compounds deepens on preservation of their stability, bioactivity, and bioavailability [54]. Hydroxycinnamic acids include caffeic acid, ferulic acid, and *p*-coumaric acids. They are sensitive to oxidation and can be rapidly oxidized, resulting in loss of bioactivity and bioavailability [55–58].

Complexation of rutin with β-CD improved its antioxidant activity-based protection of cells against oxidative stress [59].

Complexation of rutin with HP-β-CD or HP-γ-CD improved the stability of rutin and increased its antioxidant capacity, in which HP groups of CD contributed to interactions between the rutin A ring and the cavity of the CD [60]. Carotenoids have been encapsulated by CDs, and this also enhances their antioxidative activity. It is reported that methyl-β-CDs including carotenoids, such as zeaxanthin, lutein, lycopene, and β-carotene, can be used for cell supplementation [61]. After inclusion-complex formation, the stability of the methyl-β-CD–carotenoid complexes is higher than that of the free carotenoids. Nevertheless, only water-soluble inclusion complexes of the dietary carotenoid lycopene with α-CD and β-CD have been prepared for preventing oxygen-induced degradation.

Astaxanthin is a well-studied carotenoid, and its antioxidant activity can be several times higher than that of β-carotene and vitamin E [62]. However, it is a highly unsaturated molecule that can be easily decomposed by light and oxygen, and decomposition can diminish its antioxidant properties. Both β-CD and HP-β-CD can include astaxanthin to form compounds that are resistant to oxygen.

Rapid oxidation of free fatty acids in foods often results in deterioration of food quality. CDs can remove free fatty acids, which protects them against automatic oxidation [63]. By forming inclusion complexes with cyclodextrins, vitamins such as retinol and tocotrienol are effectively protected against deterioration [64–67]. Compared with α-CD and β-CD, γ-CD is particularly suited to complexing with fat-soluble vitamins, polyunsaturated fatty acids and sensitive color compounds, and bulky guests such as vitamin D2; tocopherol; omega-3, 6, and 9 fatty acids; lycopene; lutein; and anthocyanin. The large cavity size of γ-CD can achieve a nearly complete complexation and provide better stabilization against autoxidation during storage [68]. Furthermore, when used in food formulations, γ-CD can stabilize and protect certain sensitive color compounds and unique flavors during rigorous food processing procedures such as freezing, thawing, and microwaving, which preserves their quality and quantity to a greater extent and for a longer period compared with α-CD and β-CD [16, 69, 70].

3.2.2.2 *CDs protect against light-induced decomposition*

The formation of inclusion complexes between cyclodextrins and many plant bioactive compounds and flavors serves as a promising pathway for food product development. The application of CDs as carriers of these natural components is a viable choice for protecting them from degradation by light, preserving their structural integrity, and improving their biological properties.

Citral is present in the oils of several plants, including lemon myrtle, *Litsea citrata*, *Litsea cubeba*, lemongrass, lemon tea tree, *Ocimum gratissimum*, *Lindera citriodora*, petitgrain, orange, and various other fresh citrus odors. Citral is widely used in food flavors and fragrances but is sensitive to light, especially UV irradiation, and it can be cyclized to form photocitral [52]. The resulting *p*-cymene and other cyclic monoterpenes change the taste of citrus juices. Cyclodextrins can entrap citral in the hydrophobic cavity to protect against light-induced decomposition [24]. The protective effect of CD complexation against light-induced decomposition of aldehyde-type flavors was reported by Szente and Szejtli in 1987 [52]. The results showed that the protective effect against light-induced alteration in aqueous solutions and suspensions was only 15–25% of that obtained in the solid state, due to the partial release of entrapped aromatic groups of flavor compounds upon contact with water, followed by the dissociation of inclusion complexes in aqueous systems.

The stability of rutin was improved by complexation with CDs, and HP groups enhanced the stability of interactions in the complex. NMR analysis showed that the rutin A ring is inside the cavity of HP-β-CD. The inclusion complex protects rutin from thermal- and radiation-induced degradation and increases the phenolic antioxidant capacity. Quercetin is a flavonol with antibacterial, antioxidant, and antitumor properties, but it is sensitive to light. Cyclodextrins were able to improve the solubility and photostability of quercetin [71–73]. Jullian *et al.* and Calabrò *et al.* have reported 1:1 stoichiometric complexes, and Sri *et al.* described a 1:2 complex. It is possible to find different interactions between CDs and guest molecules at the same equilibrium concentrations. Moreover, the K values reported in these studies rise with increasing temperature. Carlotti *et al.* also reduced the photodegradation ratio of quercetin by complexation with M-β-CD [74]. They claimed that because quercetin was in an apolar environment inside the CD cavity, the photolytic reaction was reduced, and the amount of light capable of reaching the flavonol was lower, since it had to penetrate the CD molecule.

Ferulic acid is commonly used for preventing UV light-induced skin tumors, but it has low stability under thermal and physical stress. Inclusion complexes between ferulic acid and α-CD were studied by Anselmi *et al.* [75]. The formation of these inclusion complexes lowered penetration on the skin, which enhanced the skin protection against UV damage, since ferulic acid remained at the skin surface [76]. Protection against decomposition caused by irradiation with UV light was enhanced by complexation of ferulic acid with HP-β-CD [77]. Resveratrol is a polyphenolic with a high level of therapeutic potential due to its anticarcinogenic and antioxidant activities [78]. However, it is extremely affected by exposure to UV irradiation, which decreases its bioactivity. The use of CDs to protect resveratrol and increase its solubility, stability, and bioactivity has been described in several studies [79–84].

3.2.2.3 *CDs protect against heat-induced changes*

Cyclodextrins serve to protect food components which are sensitive to heat. Generally, complex formation between hosts (CDs) and some guests is dependent on the size and/or shape of hydrophobic cavities. Cavity diameters are 0.47–0.53, 0.6–0.65, and 0.75–0.83 nm for α-CD, β-CD, and γ-CD, respectively. The exterior diameters are 1.46 ± 0.04, 1.54 ± 0.04, and 1.75 ± 0.04 nm for these common cyclodextrins [85]. Li *et al.* investigated the effect of cyclodextrins and their derivatives on the thermal stability of the resveratrol [80]. Based on their thermal analysis, the formation between CDs and resveratrol appeared to be favored by temperature, which was interpreted as the negative enthalpy. γ-CD was able to interact better with resveratrol molecules due to its geometric dimensions, since it has the largest hydrophobic cavity of the three native cyclodextrins and is better able to encapsulate resveratrol in this larger hydrophobic cavity. A similar study used α-CD, β-CD, HP-β-CD, and DM-β-CD to increase the concentration and stability of resveratrol in solution. It was confirmed that part of the resveratrol molecule was entrapped in the hydrophobic cavity, and the exterior size/shape of HP-β-CD enhanced the dispersion of resveratrol, which gave rise to the observed increase in its thermal stability [86].

The thermal stability of anthocyanin extract isolated from the dry calyces of *Hibiscus sabdariffa* L. was studied over a temperature range of 60−90°C in aqueous solution in the presence or absence of β-CD. The thermal degradation of anthocyanins followed first-order reaction kinetics. In the presence of β-CD, anthocyanins were degraded at a slower rate, evidently due to their complexation with β-CD, and the activation energy was the same. The formation of complexes in solution was confirmed by NMR studies of β-CD solutions in the presence of the extract. Moreover, differential scanning calorimetry (DSC) revealed that the inclusion complex in the solid state was more stable against oxidation than the free extract, and the complex remained intact at temperatures of 100−250°C, whereas the free extract was oxidized at these temperatures. These results clearly indicate that the presence of β-CD improves the thermal stability of nutraceutical antioxidants present in *H. sabdariffa* L. extract, both in solution and in the solid state [87].

Volatile flavors and essential oils in the solid, dry state can be included by cyclodextrins, which possess a remarkable resistance toward thermal energy. Flavor retention was improved when employing cyclodextrins compared with traditional formulations [88]. Freely adsorbed and cyclodextrin-entrapped forms of natural essential oils were compared, and encapsulated complexes protected volatile active ingredients against evaporation. The loss of volatiles from garlic, onion, dill, and coffee occurs at high temperatures in a nitrogen atmosphere. Release profiles of essential oils complexed with β-CD indicate fewer losses compared with compounds adsorbed onto lactose [52]. Similar results were obtained following isothermic heat treatment (accelerated stability tests) at 60°C for 14 days. Again,

β-CD-complexed natural garlic oil and lemon oil were retained at levels higher than those observed with lactose-adsorbed compounds.

Nicoleta *et al.* investigated the thermal stability of linoleic acid complexed with α- and β-cyclodextrins. Bionanoparticles were obtained in solution and characterized by DSC and TEM [89]. Pure linoleic acid, corresponding thermally (50–150°C) degraded raw linoleic acid samples, and samples recovered from complexes were analyzed by gas chromatography-mass spectrometry after conversion to methyl esters. The main degradation products (for the thermally degraded raw samples) were aldehydes, epoxy and dihydroxy derivatives, and isomers of linoleic acid. A high thermal stability of nanoparticles was observed, especially for the linoleic acid–α–CD complex, which contained a relative fatty acid concentration above 98% following temperature degradation at 50 and 100°C. A lower concentration of 92% was measured in the case of the linoleic acid–β-CD complex, but following temperature degradation of 150°C, linoleic acid was partially converted to more stable geometrical isomers [89]. The thermal stability of proteins was investigated in the presence of cyclodextrins and their derivatives. Since unfolding of proteins normally involves exposure of burred hydrophobic side chains, the binding of cyclodextrins to these exposed residues could destabilize the native conformation by shifting the equilibrium in favor of the unfolded polypeptide chain.

3.2.2.4 *CDs improve the shelf life of food products*

There is growing interest in the use of CDs to improve the shelf life of food products, due to the following advantages: (a) most natural and synthetic flavors show a remarkably extended shelf life during long-term storage; (b) retrogradation of starch is inhibited by CDs, which prolongs shelf life; (c) CDs can control the release of guests included in packing materials to ensure high food quality and safety [90]. Included guests generally contain antiseptic, conserving, and antimicrobial agents.

Based on the formation of flavor compound–CD complexes, particular flavors released from foods can be controlled during storage. Szente *et al.* reported the storage of natural and synthetic flavors in complexes under normal conditions for 14 years using GC analysis [53]. Comparative TLC and GC also demonstrated that molecular encapsulation of flavors can markedly improve their long-term stability. The stability of inclusion complexes depends on the structure, polarity, and geometry of the different flavor substances. The most powerful protection has been observed with terpenoids, phenylpropanes, and alkylsulfide-type flavors, while phenolic structures are protected to a much lower extent.

The inhibitory effect of CDs on the short-term retrogradation of rice starch has been reported [91, 92], and retrogradation of rice amylose is reduced more by HP-β-CD than by β-CD. The observed intermediate crystalline pattern indicates that HP-β-CD preferentially interacts with amylose to form a potential amylose–HP-β-CD complex. Furthermore, the degree of starch retrogradation is significantly lower with β-CD than with glycerol monostearate (GMS), suggesting a different type of interaction may be responsible for the superior inhibitory activity of β-CD [91]. Nevertheless, β-CD cannot be included in starch molecules due to its larger size, and it disrupts amylose–lipid inclusion complexes [93] and may form an amylose–β-CD noninclusion complex or β-CD–lipid inclusion complex. β-CD significantly reduces the retrogradation of starch with a high amylose content but does not affect the retrogradation of waxy starch with lower amylose levels [91]. This indicates a distinct interaction between β-CD and amylose molecules. A recent study on the effect of β-CD on the recrystallization kinetics of rice amylose reported that β-CD reduces the constant rate (k) and increases the Avrami exponent (n) to a value close to 1.

These findings suggest that β-CD modifies the nucleation of amylose recrystallization to form rod-like sporadic nuclei. The occurrence of a novel crystallite further confirms the formation of an amylose–β-CD noninclusion complex. It has also been reported that β-CD inhibits the long-term retrogradation of rice starch [94, 95]. This inhibitory effect is mainly attributed to the interaction between β-CD and the amylose–lipid complex [96]. Thus, an amylose–β-CD–lipid complex presumably forms, suggesting that β-CD not only disrupts the amylose–lipid complex to generate an amylose–β-CD noninclusion complex [92, 94] but also includes lipids to form an amylose–β-CD inclusion complex in starch–β-CD systems. The resulting short, fat molecules are difficult to crystallize in a starch gel [97], which explains the inhibitory effect of β-CD on the long-term retrogradation. Bread staling can shorten the shelf-life of bread products, leading to economic losses. This has encouraged researchers to develop new techniques to control or retard the staling process. A recent report describes a novel approach to inhibit bread staling by the addition of β-CDs [98], which reduces the hardness and improves the cohesiveness and springiness of bread during storage. This is explained by a change from instantaneous nucleation to rod-like crystals. This transformation may be induced by a change in the electrostatic properties relative to the surface activity of the surrounding β-CD molecules or starch molecules. The crystalline patterns in bread crust and bread crumb are also affected by β-CDs, which result in the development of 4.4 and 6.8 Å peaks, and a weak peak at ∼5.1 Å, and an increase in the relative crystallinity of 10.2% compared with fresh crust. This indicates that a V-pattern is formed in the presence of β-CDs. Additionally, β-CDs retard the transformation from a V-type to an A-type crystalline pattern in aging crust by forming a V+A pattern crystallite. For fresh crumb, an A-type pattern is pronounced, and its transformation to a B-type pattern is slowed by the formation of an intermediate B+V crystallite formed due to the presence of β-CDs [98].

Some antiseptic, conserving, and antimicrobial agents that serve as major active ingredients in packing materials can be encapsulated by CDs. Ayala-Zavala *et al.* proposed an advantageous approach to deliver antimicrobial compounds using CDs as carriers [99]. CDs can contribute to antimicrobial delivery systems due to their ability to generate inclusion complexes with antimicrobial and antioxidant agents and enable their slow release when humidity levels increase in the headspace of packaged fresh-cut fruits and vegetables. The released antimicrobial molecules can then effectively protect products against microbial growth and oxidation induced by light and oxygen. For example, Qian *et al.* used β-CDs to complex with cinnamaldehyde to form an antimicrobial product using a thermally sealed control method [100]. Furthermore, in order to incorporate antimicrobial products into packing materials, β-CDs are first grafted onto cellulose fibers using polyacrylic acid (PAA) as a cross-linking agent, and the resultant fiber–β-CD complex can encapsulate the antimicrobial cinnamaldehyde [101] to produce an antimicrobial film that enhances the shelf life of bread to a greater extent than does the pure β-CD–cinnamaldehyde inclusion complex. This suggests that fiber–β-CD–cinnamaldehyde-based films are promising active food packing materials. Allyl isothiocyanate (AITC) is a potential natural antimicrobial agent for the preservation of foods including plant seeds, bread, cooked meat, fresh produce, and cheese [102]. Nevertheless, there are several disadvantages to using AITC directly in food systems due to its volatility and strong odor that may affect the taste. The use of AITC in food packaging materials is also limited by its physical state, which is an oil. It was therefore suggested that AITC could be encapsulated by CDs before their application, and encapsulation did indeed result in CD–AITC inclusion complexes with an improved ability to mask strong odors and prolong the antimicrobial effect. A series of AITC-based inclusion complexes based on α-CD and β-CD were prepared using a coprecipitation method that has the advantage of easy observation of complex formation [103]. The molar ratio of guest to CD is considered one of the main factors affecting complex preparation, along with the concentration of CD, the reaction temperature, the addition of ethanol, and washing with water. The release of AITC from α-CD–AITC and β-CD–AITC inclusion complexes has also been investigated at relative humidities (RHs) of 98%, 75%, and 50% [103], and the amount of complexed AITC is rapidly reduced during the initial stages of storage, and the release then becomes very slow. The release rate is dependent not only on the relative humidity but also on the content of AITC occluded in the inclusion complexes. The addition of the β-CD–AITC inclusion complex in chilled beef slices has been investigated [104], and 4 μL/g AITC and its β-CD-based inclusion complex can effectively reduce the number of *E. coli* and mesophilic aerobic bacteria. However, the addition of the β-CD-based inclusion complex results in beef slices that are whiter in color, and it can therefore only be added if this is not a problematic outcome.

3.3 Special applications of CDs in foods

In recent years, CDs and their derivatives have also been employed as "building blocks" to develop a wide variety of polymeric networks and assemblies. Besides their applications in separation sciences such as chiral liquid chromatography, capillary electrophoresis, and chiral gas chromatography [105], polymeric networks have been exploited in many other fields. Hydrogels, nano/microparticles, and micelles have been prepared using superpolymers and CDs, either through chemical cross-linking or physical assembly, and used in food and pharmaceutical applications [106].

The synthesis of polymeric hydrogels containing CDs via frontal polymerization (FP) was reported for the first time by Nuvoli *et al.* [107]. The results showed that a significant increase in the glass transition temperature and compression modulus was achieved when greater than 1 mol% of CDs was added. This study demonstrated that FP is a convenient technique for producing CD-containing hydrogels, in which the type and amount of CDs can be suitably modulated to tune the polymer properties to produce gels with the desired properties for applications in food and flavor chemistry.

Eugenol (4-allyl-2-methoxyphenol, EG), a major phenolic component from clove oil (*Eugenia caryophyllata*, Myrtaceae), has several biological activities including anti-inflammatory, analgesic, antioxidant, and antibacterial properties. However, light sensitivity and poor water solubility are major disadvantages of EG that limit its practical uses. A new water-soluble β-CD-grafted chitosan derivative (QCD-g-CS, Fig. 3.1) was synthesized under acidic conditions and high temperatures in the presence of dimethyl formamide (DMF), and an inclusion complex with EG was prepared and confirmed by FTIR. EG was found to be included both in the β-CD hydrophobic cavity of QCD-g-CS and in the hydrophobic core of QCD-g-CS self-aggregates, resulting in varying degrees of EG entrapment. The degree of QCD substitution in QCD-g-CS plays an important role in determining

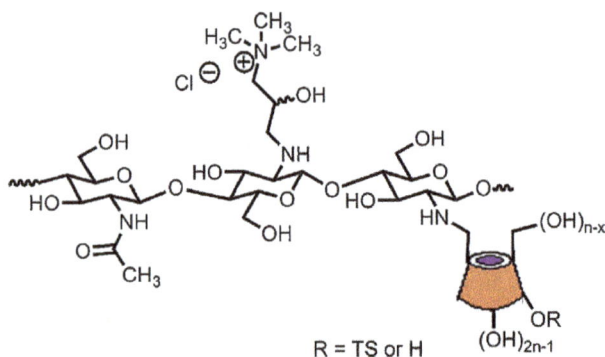

Fig. 3.1.   Chemical structure of a QCD-g-CS derivative.

their physical properties, due to steric hindrance. QCD-g-CS showed excellent mucoadhesion, and the inclusion complex between QCD-g-CS and EG displayed higher antimicrobial activity than the native QCD-g-CS [108].

Dietary fiber is an important component of a healthy human diet. It can improve gastrointestinal health and may reduce the risk of coronary heart disease and other related ailments.

Over recent decades, CDs and their ability to control the release of active components from active or controlled packaging have received increasing attention. Many publications have reported the successful application of CDs in polymer syntheses in aqueous solution, and the properties of CD-containing films have been reviewed. The preparation of CD-containing polymers has been studied by those working in the alimentary field, including complexes containing antimicrobial and antioxidizing compounds. In 2011, an active food packaging system based on the incorporation of agents into polymeric package walls was reported by López-de-Dicastillo *et al.* that can purposely release or retain compounds to maintain or even increase food quality [109]. The researchers developed polyvinyl alcohol (PVOH)–β-CD composite films that can be applied to reduce the amount of undesirable components such as cholesterol in foods through active retention of the compounds in the package walls during storage. CDs were added to PVOH at 1:1 ratio and cross-linked with glyoxal under acidic conditions to reduce the water-soluble character. Three different cross-linking procedures were carried out in the study: cross-linking of the polymer–polysaccharide mixture in solution and film casting (PVOH.β-CD*); cross-linking of the polymer, addition of β-CD, and casting of the mixture (PVOH*.CD); and casting of a PVOH film, addition of a β-CD/glyoxal solution onto the film, and cross-linking during drying (PVOH.CD*). The PVOH*.CD and PVOH.CD* films displayed the best physical characteristics with the lowest release values and highest barrier properties. The prepared films were tested as potential cholesterol-scavenging materials, and a significant reduction in the cholesterol concentration in milk samples was observed. The research group also reported the immobilization of β-CD in ethylene-vinyl alcohol copolymers [110]. In the study, β-CD was successfully immobilized in an ethylene–vinyl alcohol copolymer with a 44% molar percentage of ethylene (EVOH44) using regular extrusion with glycerol as an adjuvant. Films with 10%, 20%, and 30% β-CD were flexible and transparent. The addition of the agent slightly increased the glass transition temperature and crystallinity percentage of the polymer. This indicated the introduction of some fragility and promotion of nucleation. Compared with the pure polymeric material, permeability to water vapor, oxygen, and carbon dioxide was increased with the addition of β-CD, due to the presence of discontinuities in the matrix and to the internal cavity of the oligosaccharide. Moreover, the CO_2/O_2 permselectivity was also increased with the addition of β-CD. Materials containing β-CD preferentially absorbed apolar compounds such as terpenes. The authors suggested that this characteristic could be useful in active packaging applications for preferentially retaining undesirable apolar food components such as hexanal or cholesterol.

Lizundia *et al.* developed a poly(L-lactide)/branched β-CD blend in an attempt to obtain new biocompatible and biodegradable materials of pharmaceutical, biomedical, and food industry interests. The chosen material was naturally available and had tunable functional properties. In the study, ionic branched β-CD was obtained by polycondensation of the β-CD monomer and blending with a commercially available PLLA. A single glass transition was observed for the blended material, and the glass transition temperature was decreased from 57.4°C for neat PLLA to 31.1°C for the 50–50 blend. DSC results showed that β-CD speeded up the crystallization process of the PLLA phase due to the increased mobility of PLLA molecules induced by the presence of β-CD molecules. FTIR results demonstrated the occurrence of specific interactions in PLLA–β-CD blends, while morphological studies showed single phase morphology over the entire composition, demonstrating good miscibility between the individual components. Analysis of the thermal properties of blends revealed that the addition of β-CD continuously reduced the thermal stability of the blends. This work demonstrated that PLLA could be successfully utilized as a supporting matrix for the development of materials in which introduced CD derivatives can retain and release specific organic molecules, generating materials that are especially suitable for active packaging applications [111].

In 2007, Fenyvesi *et al.* described the pilot scale production of CD-containing polyvinyl chloride (PVC) and polyethylene (PE) films [112]. 0%, 1.0%, 1.5%, and 2.0% β-CD were added and their permeation properties were characterized using model flavors including carvone, vanillin, and diacetyl. The results showed that complexes containing carvone and vanillin could penetrate through the films containing β-CD, while those containing diacetyl, which has a poor complex-forming ability with β-CD, could not. The permeation of the included ingredient was found to be dependent on the CD concentration and the thickness of the film. In addition, the presence of β-CD in the film results in a slower leaching of plasticizers from the polymer matrix.

A relatively small internal cavity and high resistance to enzymatic hydrolysis make α-CD well-suited to applications in many fields, especially in the food industry. However, the market share of α-CD is currently much lower than that of β-CD due to its high price and low production yield. Vast efforts have been undertaken to optimize α-CD production processes by modifying the properties of cyclodextrin glycosyltransferases (CGTases). By increasing the yield, the application of α-CD will be increased significantly over the coming years [113]. α-CDs have recently been approved in Europe as novel food ingredients (functional foods) for use as prebiotics. This term is used to describe nondigestible carbohydrates or dietary fiber compounds that support the gut flora in the large intestine. Dietary fiber is a medically important component of the human diet. It can protect the health of the gastrointestinal tract and may reduce the risk of coronary heart disease and other related ailments. CDs could therefore potentially increase the intake of dietary fiber, which is essential for intestinal health.

Soluble fiber can dissolve in water to form a hydrogel that may help reduce cholesterol and blood glucose levels. α-CD is a natural, soluble dietary fiber component, and α-CD is used increasingly in the food industry for this purpose. Although triglycerides are the main form of fat in the body and are essential to human life, high triglyceride levels raise the risk of heart diseases [114]. Among the known dietary fiber components, α-CD has the unique ability to bind several times its own weight in fat molecules. Moreover, α-CD–fat complexes are resistant to lipolytic hydrolysis by typical lipases. As a result, α-CD reduces the absorption and bioavailability of dietary fats, and it can therefore be used as a weight loss supplement. Indeed, animal research has shown that α-CD significantly reduces weight gain in rats [115]. Most importantly, α-CD does not cause fat malabsorption which can lead to gastrointestinal side effects associated with weight loss products since α-CD-fat complexes are not accessible to the human gut flora.

Besides weight control, α-CD can also provide other health benefits. α-CD can significantly reduce blood lipid level and increase adiponectin level in obese type 2 diabetic individuals. This may potentially be helpful for the treatment of type 2 diabetes. α-CD can also reduce the glycemic impact of carbohydrate foods.

As a soluble dietary fiber component, α-CD may not only be used as additive in beverages but can also be added to other foods. α-CD can be applied in bakery products and other finished foods because they remain stable at high temperatures. Moreover, the typical discoloration and formation of flavors in the Maillard reaction do not occur since α-CD contains no reducing sugars. By comparison, glucose and maltodextrin lead to significant discoloration upon reaction with the amino acid glycine. Additionally, unlike other fibers such as inulin, α-CD has a very low hygroscopicity, which ensures that bakery products remain fresh and crisp during storage.

A metal-organic framework (MOF) is an extensive class of porous crystals in which organic struts link metal-containing clusters. The ability to easily modify the structures and functions of MOFs has led to numerous applications in gas adsorption, storage of clean gas fuels, catalysis, separations, and drug/nutrient delivery. An MOF derived from edible natural ingredients was reported in 2010. A food-grade γ-CD and KOH were used to form the CD-MOF framework by combining 1.0 equiv. of γ-CD with 8.0 equiv. of KOH in aqueous solution, followed by vapor diffusion of MeOH into the solution over 2–7 days, resulting in colorless, single cubic crystals in approximately 70% yield. The crystal structure of the CD-MOF is shown in Fig. 3.2 [116]. Following this, in 2016, curcumin was successfully encapsulated in the CD-MOF by Moussa *et al.* without altering the crystallinity. It was found that the interaction between

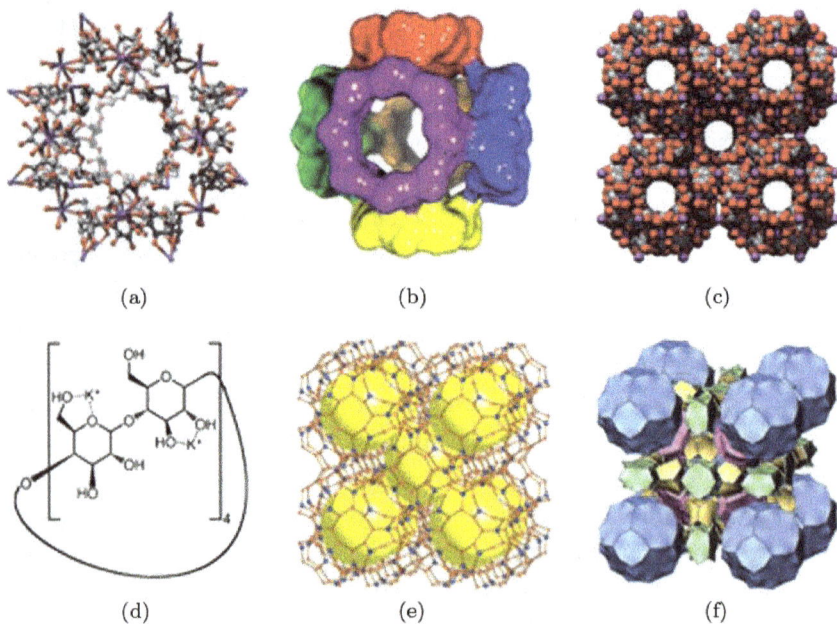

(a)   (b)   (c)

(d)   (e)   (f)

Fig. 3.2.   Crystal structure of CD-MOF.

curcumin and CD-MOFs was strong due to hydrogen bond-type interactions between the OH group of the γ-CD of CD-MOFs and the phenolic hydroxyl group of curcumin. Interestingly, a unique complex between curcumin, γ-CD, and potassium cations was formed after dissolving the curcumin-loaded CD-MOF crystals in water. The stability of curcumin in this complex was enhanced at least threefold compared with free curcumin and curcumin:γ-CD at pH 11.5. These results indicate that these complexes offer a promising benign system that could be used to store and stabilize organic compounds in food applications.

References

1. Chernykh EV, Brichkin SB. (2010). Supramolecular complexes based on cyclodextrins. *High Energy Chemistry*, 44(2), pp. 83–100.
2. Brewster ME, Loftsson T. (2007). Cyclodextrins as pharmaceutical solubilizers. *Advanced Drug Delivery Reviews*, 59(7), pp. 645–666.
3. Kurkov SV, Loftsson T. (2013). Cyclodextrins. *International Journal of Pharmaceutics*, 453(1), pp. 167–180.
4. Szente L, Szejtli J. (2004). Cyclodextrins as food ingredients. *Trends in Food Science & Technology*, 15(3–4), pp. 137–142.
5. Szejtli J. (1988). *Cyclodextrin Technology*. Dordrecht, The Netherlands: Kluwer Academic, pp. 307–328.
6. Szente L, Szejtli J. (2004). Cyclodextrins as food ingredients. *Trends in Food Science & Technology*, 15, pp. 137–142.
7. Li Z, Chen S, Gu Z, Chen J, Wu J. (2014). Alpha-cyclodextrin: enzymatic production and food applications. *Trends in Food Science & Technology*, 35, pp. 151–160.
8. WHO. (2002). Safety evaluation of certain food additives and contaminants. In *WHO Food Additives Series*, Vol. 48. Available from: www.inchem.org/documents/jecfa/jecmono/v48je10.htm. Last accessed 10 July 2013.
9. Jin Z-Y. (2013). *Cyclodextrin Chemistry: Preparation and Application*. Singapore: World Scientific Publishing Co Pte Ltd., pp. 215–223.
10. Lubbers S, Landy P, Voilley A. (1998). Retention and release of aroma compounds in food containing proteins. *Food Technology Chicago*, 52, pp. 68–74.
11. Astray G, Gonzalez-Barreiro C, Mejuto JC, Rial-Otero R, Simal-Gándara J. (2009). A review on the use of cyclodextrins in foods. *Food Hydrocolloid*, 23, pp. 1631–1640.
12. Martin Del Valle EM. (2004). Cyclodextrins and their uses: a review. *Process Biochemistry*, 39, pp. 1033–1046.
13. Lindner K. (1982). Using cyclodextrin aroma complexes in the catering. *Nahrung*, 26, pp. 675–680.

14. Szetjli J. (1998). Introduction and general overview of cyclodextrin chemistry. *Chemical Reviews*, 98, pp. 1743–1753.
15. Hedges RA. (1998). Industrial applications of cyclodextrins. *Chemical Reviews*, 98, pp. 2035–2044.
16. Muñoz-Botella S, del Castillo B, Martyn MA. (1995). Cyclodextrin properties and applications of inclusion complex formation. *Ars Pharmaceutica*, 36, pp. 187–198.
17. Parrish MA. (1998). *Cyclodextrins: A Review*. Newcastle upon Tyne: Sterling Organics.
18. Sojo MM, Nunez-Delicado E, Garcia-Carmona F, Sanchez-Ferrer A. (1999). Cyclodextrins as activator and inhibitor of latent banana pulp polyphenol oxidase. *Journal of Agricultural and Food Chemistry*, 47, pp. 518–523.
19. Sung H. (1997). Composition for ginger preservation. Republic of Korea KR 9,707,148.
20. Sumiyoshi H. (1999). Utilisation of inclusion complexes with plant components for foods. *Nippon Shokuhin Shinsozai Kenkyukaishi*, 2, pp. 109–114.
21. Fujishima N, Kusaka K, Umino T, Urushinata T, Terumi K. (2001). Flour based foods containing highly branched cyclodextrins. Japanese Patent JP 136,898.
22. Takeshita K, Urata T. (2001). Antimicrobial food preservatives containing CD inclusion complexes. Japanese Patent JP 29,054.
23. Mabuchi N, Ngoa M. (2001). Controlled release powdered flavor preparations and confectioneries containing preparations. Japanese Patent JP 128,638.
24. Szejtli J, Szente L, Banky-Elod E. (1979). Molecular encapsulation of volatile, easily oxidizable flavor substances by cyclodextrins. *Acta Chimica Academiae Scientiarum Hungaricae*, 101, pp. 27–46.
25. Szente L, Szejtli J. (1987). Stabilization of flavors by cyclodextrins. In: Risch SJ, Reineccius G (Eds.), *Flavor Encapsulation*. American Chemical Society Symposium Series, Vol. 370, Washington, DC: American Chemical Society, pp. 148–158.
26. Shiga H, Yoshii H, Ohe H, Yasuda M, Furuta T, Kuwahara H, *et al.* (2004). Encapsulation of shiitake (*Lenthinus edodes*) flavors by spray drying. *Bioscience, Biotechnology and Biochemistry*, 68, pp. 66–71.
27. Astray G, Mejuto JC, Morales J. (2010). Factors controlling flavors binding constants to cyclodextrins and their applications in foods. *Food Research International* 43, pp. 1212–1218.
28. Szejtli J, Osa T. (1996). Comprehensive Supramolecular Chemistry, Vol. 3, 1996, Elsevier, Oxford.
29. Cravotto G, Binello A, Baranelli E, Carraro P, Trotta F. (2006). Cyclodextrins as food additives and in food processing. *Current Nutrition Food Science*, 2(4), pp. 343–350.
30. Hedges AR, McBride C. (1999). Utilization of β-cyclodextrin in food. *Cereal Foods World*, 44(10), pp. 700–704.
31. Hedges AR, Shieh WJ, Sikorski CT. (1995). Use of cyclodextrins for encapsulation in the use and treatment of food products. In: Risch SJ, Reineccius GA (Eds.), *Encapsulation and Controlled Release of Food Ingredients*, ACS Symposium Series, Vol. 590, Washington, DC: American Chemical Society, pp. 60–71.

32. Qi ZH, Hedges AR. (1995). Use of cyclodextrins for flavors. In: Ho CT, Tan CT, Tong CH (Eds.), *Flavor Technology: Physical Chemistry, Modification and Process*, ACS Symposium Series, Vol. 610, Washington, DC: American Chemical Society, pp. 231–243.

33. Samant SK, Pai JS. (1991). Cyclodextrins: new versatile food additive. *Indian Food Packer*, 45(3), pp. 55–65.

34. Cuccovia IM, Schoroter EH, Montero PM, Chaimovich H. (1978). Effect of hexadecyltrimethylammonium bromide on the thiolysis of p-nitrophenyl acetate. *Journal of Organic Chemistry*, 43, pp. 2248–2252.

35. Bhandari B, D'Arcy B, Young G. (2001). Flavor retention during high temperature short time extrusion cooking process: a review. *International Journal of Food Science and Technology*, 36, pp. 453–461.

36. Jouquand C, Ducruet V, Giampaoli P. (2004). Partition coefficients of aroma compounds in polysaccharide solutions by the phase ratio variation method. *Food Chemistry*, 85, pp. 467–474.

37. Pagington JS. (1987). β-cyclodextrin: the success of molecular inclusion. *Chemistry in Britain*, 23, pp. 455–458.

38. Binello A, Cravotto G, Nano GM, Spagliardi P. (2004). Synthesis of chitosan-cyclodextrin adducts and evaluation of their bitter-masking properties. *Flavor Fragrance Journal*, 19(5), pp. 394–400.

39. Binello A, Robaldo B, Barge A, Cavalli R, Cravotto G. (2008). Synthesis of cyclodextrin-based polymers and their use as debittering agents. *Journal Applied Polymer Science*, 107, pp. 2549–2557.

40. Shaw PE, Tatum JH, Wilson CW. (1984). Improved flavor of navel orange and grapefruit juices by removal of bitter components with β-cyclodextrin polymer. *Journal of Agricultural and Food Chemistry*, 32, pp. 832–836.

41. Singh M, Sharma R, Banerjee UC. (2002). Biotechnological applications of cyclodextrins. *Biotechnology Advances*, 20, pp. 341–359.

42. Suzuki J. (1975). Japan Kokai, JP 7 569 100.

43. Kuwabara N, Takaku H, Oku S, Kopure Y. (1988). Japan Kokai, JP 88 267 246.

44. Sakakibara S, Sugisawa K, Matsui F, Sengoku K. (1985). Japanese Patent JP 851248 075.

45. Takeda Chem. Ind. Ltd. (1981). Japan Kokai, JP 81 127 058.

46. López-Nicolas JM, García-Carmona F. (2007). Use of cyclodextrins as secondary antioxidants to improve the color of fresh pear juice. *Journal of Agricultural and Food Chemistry*, 55, pp. 6330–6338.

47. López-Nicolas JM, Nunez-Delicado E, Sánchez-Ferrer A, García-Carmona F. (2007). Kinetic model of apple juice enzymatic browning in the presence of cyclodextrins: the use of maltosyl-β-cyclodextrin as secondary antioxidant. *Food Chemistry*, 101, pp. 1164–1171.

48. López-Nicolas JM, Pérez-López AJ, Carbonell-Barrachina A, García-Carmona F. (2007a). Use of natural and modified cyclodextrins as inhibiting agents of peach juice enzymatic browning. *Journal of Agricultural and Food Chemistry*, 55, pp. 5312–5319.

49. López-Nicolas JM, Pérez-López AJ, Carbonell-Barrachina A, García-Carmona F. (2007b). Kinetic study of the activation of banana juice enzymatic browning by the addition of maltosyl-β-cyclodextrin. *Journal of Agricultural and Food Chemistry*, 55, pp. 9655–9662.

50. Sojo MM, Nuñez-Delicado E, García-Carmona F, *et al.* (1999). Cyclodextrins as activator and inhibitor of latent banana pulp polyphenol oxidase. *Journal of Agricultural and Food Chemistry*, 47(2), pp. 518–523.

51. Szente L, Szejtli J. (1987). Stabilization of flavors by cyclodextrins. In *Abstracts of Papers of the American Chemical Society*, Vol. 194. Washington, DC: American Chemical Society, p. 96.

52. Szente L, Szejtli J. (2004). Cyclodextrins as food ingredients. *Trends in Food Science & Technology*, 15, pp. 137–142.

53. Fang Z, Bhandari B. (2010). Encapsulation of polyphenols: a review. *Trends in Food Science & Technology*, 21(10), pp. 510–523.

54. Crozier A, Jaganath IB, Clifford MN. (2009). Dietary phenolics: chemistry, bioavailability and effects on health. *Natural Product Reports*, 26(8), pp. 1001–1043.

55. Munin A, Edwards-Lévy F. (2011). Encapsulation of natural polyphenolic compounds: a review. *Pharmaceutics*, 3(4), pp. 793–829.

56. Nichols JA, Katiyar SK. (2010). Skin photoprotection by natural polyphenols: anti-inflammatory, antioxidant and DNA repair mechanisms. *Archives of Dermatological Research*, 302(2), pp. 71–83.

57. Yang LJ, Chen W, Ma SX, *et al.* (2011). Host–guest system of taxifolin and native cyclodextrin or its derivative: preparation, characterization, inclusion mode, and solubilization. *Carbohydrate Polymers*, 85(3), pp. 629–637.

58. Calabro ML, Tommasini S, Donato P, *et al.* (2005). The rutin/β-cyclodextrin interactions in fully aqueous solution: spectroscopic studies and biological assays. *Journal of Pharmaceutical and Biomedical Analysis*, 36(5), pp. 1019–1027.

59. Nguyen TA, Liu B, Zhao J, *et al.* (2013). An investigation into the supramolecular structure, solubility, stability and antioxidant activity of rutin/cyclodextrin inclusion complex. *Food Chemistry*, 136(1), pp. 186–192.

60. Pfitzner I, Francz PI, Biesalski HK. (2000). Carotenoid: methyl-β-cyclodextrin formulations: an improved method for supplementation of cultured cells. *Biochimica et Biophysica Acta(BBA) - General Subjects*, 1474(2), pp. 163–168.

61. Yuan C, Jin Z, Xu X, *et al.* (2008). Preparation and stability of the inclusion complex of astaxanthin with hydroxypropyl-β-cyclodextrin. *Food Chemistry*, 109(2), pp. 264–268.

62. Hedges AR. (1998). Industrial applications of cyclodextrins. *Chemical Reviews*, 98(5), pp. 2035–2044.

63. Regiert M, Moldenhauer JP. (1998). Inclusion compound from gammacyclodextrin and retinol, production and application. Eur Patent EP0867175 A, 1.

64. Regiert M, Kupka M. (2003). 2: 1-Inclusion compound from beta- or gamma-cyclodextrin and alpha-tocopherol. US Patent 20030130231A1.

65. Regiert M, Kupka M. (2010). Cosmetic composition comprising a complex of cyclodextrin and vitamin F: US Patent 7,667,057.
66. Regiert M, Wacker Chemie AG. (2005). Light-stable vitamin E by inclusion in γ-cyclodextrin. SOEFW J, 131, pp. 10–18.
67. O'Donnell CD. (2001). New Encapsulating Molecule Improves Taste[J]. Prepared Foods. Available from http://www.preparedfoods.com/articles/105847-new-encapsulating-molecule-improves-taste. Last accessed 13 September 2016.
68. Thoss M, Schwabe L, Fromming KH. (1993). Cyclodextrin-Einschlussverbindungen von Zitronen-, Orangen-, Hopfen-und Kamillenol. *Pharm Ztg Wiss*, 6, pp. 144–148.
69. Tamura H, Takada M, Yamagami A, *et al.* (1998). The color stability and antioxidative activity of an anthocyanin and γ-cyclodextrin complex. In Functional foods for disease prevention I. Fruits, vegetables, and teas. Symposium sponsored by the Division of Agricultural and Food Chemistry at the 213th National Meeting of the American Chemical Society, San Francisco, California, USA, April 13–17, 1997. American Chemical Society, pp. 157–171.
70. Calabrò ML, Tommasini S, Donato P, *et al.* (2004). Effects of α- and β-cyclodextrin complexation on the physico-chemical properties and antioxidant activity of some 3-hydroxyflavones. *Journal of Pharmaceutical and Biomedical Analysis*, 35(2), pp. 365–377.
71. Jullian C, Moyano L, Yanez C, *et al.* (2007). Complexation of quercetin with three kinds of cyclodextrins: an antioxidant study. *Spectrochimica Acta Part A: Molecular and Biomolecular Spectroscopy*, 67(1), pp. 230–234.
72. Sri KV, Kondaiah A, Ratna JV, *et al.* (2007). Preparation and characterization of quercetin and rutin cyclodextrin inclusion complexes. *Drug Development and Industrial Pharmacy*, 33(3), pp. 245–253.
73. Carlotti ME, Sapino S, Ugazio E, *et al.* (2011). On the complexation of quercetin with methyl-β-cyclodextrin: photostability and antioxidant studies. *Journal of Inclusion Phenomena and Macrocyclic Chemistry*, 70(1–2), pp. 81–90.
74. Anselmi C, Centini M, Maggiore M, *et al.* (2008). Non-covalent inclusion of ferulic acid with α-cyclodextrin improves photo-stability and delivery: NMR and modeling studies. *Journal of Pharmaceutical and Biomedical Analysis*, 46(4), pp. 645–652.
75. Monti D, Tampucci S, Chetoni P, *et al.* (2011). Permeation and distribution of ferulic acid and its α-cyclodextrin complex from different formulations in hairless rat skin. AAPS PharmSciTech, 12(2), pp. 514–520.
76. Wang J, Cao Y, Sun B, *et al.* (2011). Characterisation of inclusion complex of trans-ferulic acid and hydroxypropyl-β-cyclodextrin. *Food Chemistry*, 124(3), pp. 1069–1075.
77. Sapino S, Carlotti ME, Caron G, *et al.* (2009). In silico design, photostability and biological properties of the complex resveratrol/hydroxypropyl-β-cyclodextrin. *Journal of Inclusion Phenomena and Macrocyclic Chemistry*, 63(1–2), pp. 171–180.

78. Krishnaswamy K, Orsat V, Thangavel K. (2012). Synthesis and characterization of nano-encapsulated catechin by molecular inclusion with beta-cyclodextrin. *Journal of Food Engineering*, 111(2), pp. 255–264.
79. Li H, Xu X, Liu M, *et al.* (2010). Microcalorimetric and spectrographic studies on host–guest interactions of α-, β-, γ-and Mβ-cyclodextrin with resveratrol. *Thermochimica Acta*, 510(1), pp. 168–172.
80. Lu Z, Cheng B, Hu Y, *et al.* (2009). Complexation of resveratrol with cyclodextrins: solubility and antioxidant activity. *Food Chemistry*, 113(1), pp. 17–20.
81. Lu Z, Chen R, Fu R, *et al.* (2012). Cytotoxicity and inhibition of lipid peroxidation activity of resveratrol/cyclodextrin inclusion complexes. *Journal of Inclusion Phenomena and Macrocyclic Chemistry*, 73(1–4), pp. 313–320.
82. Lucas-Abellán C, Fortea I, López-Nicolás JM, *et al.* (2007). Cyclodextrins as resveratrol carrier system. *Food Chemistry*, 104(1), pp. 39–44.
83. Sapino S, Carlotti ME, Caron G, *et al.* (2009). In silico design, photostability and biological properties of the complex resveratrol/hydroxypropyl-β-cyclodextrin. *Journal of Inclusion Phenomena and Macrocyclic Chemistry*, 63(1–2), pp. 171–180.
84. Yu B, Wang J, Zhang H, *et al.* (2011). Investigation of the interactions between the hydrophobic cavities of cyclodextrins and pullulanase. *Molecules*, 16(4), pp. 3010–3017.
85. Kumpugdee-Vollrath M, Ibold Y, Sriamornsak P. (2012). Solid state characterization of trans resveratrol complexes with different cyclodextrins. *Journal of the Asian Association of Schools of Pharmacy*, 1(2), pp. 125–136.
86. Mourtzinos I, Makris DP, Yannakopoulou K, *et al.* (2008). Thermal stability of anthocyanin extract of *Hibiscus sabdariffa* L. in the presence of β-cyclodextrin. *Journal of Agricultural and Food Chemistry*, 56(21), pp. 10303–10310.
87. Furuta T, Yoshii H, Fujimoto T, *et al.* (1996). A short-cut method for estimating l-menthol retention included in cyclodextrin during drying a single drop. In Proceedings of the Eighth International Symposium on CDs. Springer, The Netherlands, pp. 583–586.
88. Hădărugă NG, Hădărugă DI, Păunescu V, *et al.* (2006). Thermal stability of the linoleic acid/α- and β-cyclodextrin complexes. *Food Chemistry*, 99(3), pp. 500–508.
89. Wood WE. (2001). Improved aroma barrier properties in food packaging with cyclodextrins. *Paper, Film and Foil Converter*, 75(11), p. 68.
90. Tian Y, Li Y, Manthey FA, *et al.* (2009). Influence of β-cyclodextrin on the short-term retrogradation of rice starch. *Food Chemistry*, 116(1), pp. 54–58.
91. Tian Y, Li Y, Jin Z, *et al.* (2010). Comparison tests of hydroxylpropyl β-cyclodextrin (HPβ-CD) and β-cyclodextrin (β-CD) on retrogradation of rice amylose. *LWT — Food Science and Technology*, 43(3), pp. 488–491.
92. Gunaratne A, Corke H. (2007). Influence of unmodified and modified cycloheptaamylose (β-cyclodextrin) on transition parameters of amylose–lipid complex and functional properties of starch. *Carbohydrate Polymers*, 68(2), pp. 226–234.

93. Tian YQ, Jin ZY, Deng L, Zhao JW. (2008). Stability and anti-retrogratation of vitamin D3-β-cyclodextrin inclusion compound. *Journal of the Chinese Cereals and Oils Association*, 23(5), pp. 95–98.
94. Tian Y, Xu X, Li Y, *et al.* (2009). Effect of β-cyclodextrin on the long-term retrogradation of rice starch. *European Food Research and Technology*, 228(5), pp. 743–748.
95. Tian Y, Yang N, Li Y, *et al.* (2010). Potential interaction between β-cyclodextrin and amylose–lipid complex in retrograded rice starch. *Carbohydrate Polymers*, 80(2), pp. 581–584.
96. Gunning AP, Giardina TP, Faulds CB, *et al.* (2003). Surfactant-mediated solubilisation of amylose and visualisation by atomic force microscopy. *Carbohydrate Polymers*, 51(2), pp. 177–182.
97. Tian YQ, Li Y, Jin ZY, *et al.* (2009). β-cyclodextrin (β-CD): a new approach in bread staling. *Thermochimica Acta*, 489(1), pp. 22–26.
98. Ayala-Zavala JF, Del-Toro-Sánchez L, Alvarez-Parrilla E, *et al.* (2008). High relative humidity in-package of fresh-cut fruits and vegetables: advantage or disadvantage considering microbiological problems and antimicrobial delivering systems?. *Journal of Food Science*, 73(4), pp. 41–47.
99. Qian LL, Jin ZY, Deng L. (2007). Preparation of inclusion complex of cinnamaldehyde and β-cyclodextrin by sealed thermal control method. *Food and Fermentation Industries*, 33(12), pp. 13–16.
100. Qian LL, Jin ZY, Deng L, Li XH. (2008). Inclusion and release of cinnamaldehyde by β-cyclodextrin grafted on the cellulose. *Food and Fermentation Industries*, 34(2), pp. 16–20.
101. Isshiki K, Tokuoka K, Mori R, *et al.* (1992). Preliminary examination of allyl isothiocyanate vapor for food preservation. *Bioscience, Biotechnology, and Biochemistry*, 56(9), pp. 1476–1477.
102. Li X, Jin Z, Wang J. (2007). Complexation of allyl isothiocyanate by α- and β-cyclodextrin and its controlled release characteristics. *Food Chemistry*, 103(2), pp. 461–466.
103. Xuehong L, Zhengyu J. (2007). Application of allyl isothiocyanate and its complex with cyclodextrin in preservation of chilled beef slices. *Transactions of the Chinese Society of Agricultural Engineering*. doi: 10.3969/j.issn.1002-6819.2007.7.049.
104. Schneiderman E, Stalcup AM. (2000). Cyclodextrins: a versatile tool in separation science. *Journal of Chromatography B: Biomedical Sciences and Applications*, 745(1), pp. 83–102.
105. Martina K, Cravotto G. (2015). Cyclodextrin as a food additive in food processing. In: Cirillo G, Spizzirri UG, Iemma F (Eds.), *Functional Polymers in Food Science: From Technology to Biology*, Vol. 2. New York: John Wiley & Sons, pp. 267–288.
106. Nuvoli D, *et al.* (2016). Synthesis and characterization of poly(2-hydroxyethylacrylate)/β-cyclodextrin hydrogels obtained by frontal polymerization. *Carbohydrate Polymers*, 150, pp. 166–171.

107. Sajomsang W, *et al.* (2012). Water-soluble β-cyclodextrin grafted with chitosan and its inclusion complex as a mucoadhesive eugenol carrier. *Carbohydrate Polymers*, 89(2), pp. 623–631.
108. López-de-Dicastillo C, *et al.* (2011). Development of active polyvinyl alcohol/β-cyclodextrin composites to scavenge undesirable food components. *Journal of Agricultural and Food Chemistry*, 59(20), pp. 11026–11033.
109. López-De-Dicastillo C, *et al.* (2010). Immobilization of β-cyclodextrin in ethylene-vinyl alcohol copolymer for active food packaging applications. *Journal of Membrane Science*, 353(1), pp. 184–191.
110. Lizundia E, *et al.* (2016). Poly(L-lactide)/branched β-cyclodextrin blends: thermal, morphological and mechanical properties. *Carbohydrate Polymers*, 144, pp. 25–32.
111. Fenyvesi É, *et al.* (2007). Permeability and release properties of cyclodextrin-containing poly(vinyl chloride) and polyethylene films. *Journal of Inclusion Phenomena and Macrocyclic Chemistry*, 57(1–4), pp. 371–374.
112. Li Z, *et al.* (2014). Alpha-cyclodextrin: enzymatic production and food applications. *Trends in Food Science & Technology*, 35(2), pp. 151–160.
113. Kasai T, *et al.* (2013). Mortality risk of triglyceride levels in patients with coronary artery disease. *Heart*, 99(1), pp. 22–29.
114. Artiss JD, *et al.* (2006). The effects of a new soluble dietary fiber on weight gain and selected blood parameters in rats. *Metabolism*, 55(2), pp. 195–202.
115. Smaldone RA, *et al.* (2010). Metal-organic frameworks from edible natural products. *Angewandte Chemie International Edition*, 49(46), pp. 8630–8634.
116. Moussa Z, *et al.* (2016). Encapsulation of curcumin in cyclodextrin-metal organic frameworks: dissociation of loaded cyclodextrin-MOFs enhances stability of curcumin. *Food Chemistry*, 212, pp. 485–494.

4. Applications in Agriculture

Jianwei Zhao* and Shengjun Wu[†]

*School of Food Science and Technology,
Jiangnan University, Wuxi 214122, China
[†]School of Food Engineering,
HuaiHai Institute of Technology,
59 Cangwu Road, Lianyungang 222005, China

4.1 Introduction

The interior cavity of cyclodextrins (CDs) is hydrophobic due to the electron-rich glycosidic oxygen atoms, while the exterior shell is hydrophilic because of the secondary hydroxyl groups [1]. CDs can form inclusion compounds with guest molecules by adopting specific architectural conformations. In general, the formation of inclusion compounds increases the stability of guest molecules [2], controls the release of guest molecules, or enhances the bioactivity of guest molecules [3]. Moreover, CDs can be degraded by soil microorganisms through soil remediation [4]. At present, CDs are used in agriculture to control the release of fertilizers, soil amendment components, animal feed constituents, and pesticides.

4.2 Applications in fertilizers

4.2.1 *Application in controlled-release urea fertilizers*

β-CDs have been used to control the release of urea in fertilizers [5]. Urea is a neutral organic nitrogen fertilizer which is widely used due to its high nitrogen content. However, urea has shortcomings, including a low utilization ratio and high propensity for leaching. When urea is applied to soil, a proportion is absorbed by crops, some is hydrolyzed to NH_4^+ and transformed into NO_3^-, some is adsorbed by soil colloid, and the rest is volatized in the form of ammonia, resulting in nitrogen loss.

CDs are annular oligosaccharides produced from starch by CD glucose transferase-based hydrolysis. CDs can be selectively combined with various small organic molecules or ions to form host–guest inclusion complexes that can slowly release the guest molecules in response to environmental changes due to the special molecular structure of the cavity [6].

In order to reduce the rate of urea hydrolysis and improve the urea utilization rate, a complex of β-CD and urea can be prepared using coating technology [7] and urease inhibitors [8]. CDs are also used for the preparation of materials that slowly release fertilizers, and CDS themselves are then slowly degraded by soil microbes and become soil nutrients.

4.2.1.1 *Preparation of β-CD–urea inclusion complex*

β-CD was mixed with urea in a mortar at a molar ratio ranging from 0.5:1 to 4.0:1 and fully ground for 2 h and dried in a vacuum drying oven (temperature $= 40°C$). The resulting inclusion complex was analyzed, and IR spectroscopy revealed an amide asymmetric stretching vibration peak present at 3432.67 cm^{-1}, an amide symmetric stretching vibration peak present at 3318.89 cm^{-1}, and an amide stretching vibration peak present at 1641.12 cm^{-1}, comparable with the spectrum of pure β-CD. An N–H in-plane bending vibration peak at 1567.8 cm^{-1} and a peak characteristic of urea both disappeared in the IR spectrum of the β-CD–urea inclusion complex, indicating that urea was included in the β-CD cavity.

4.2.1.2 *Release rate of the β-CD–urea inclusion complex*

When the molar ratio of β-CD to urea was 1:1, the preliminary solubility rate of the β-CD–urea inclusion complex after 24 h was

14.37%, and this decreased with increasing β-CD to urea molar ratio, due to the higher inclusion rate at a higher molar ratio. The differential dissolving rate after 7 days was 2.37%, and the cumulative release rate after 28 days was 42.57%. The preliminary solubility rate, differential dissolving rate, and cumulative release rate after 28 days decreased as the urea background concentration was increased, demonstrating that the rate of urea release was slowed down in the inclusion complex [5].

4.2.2 Applications in Chinese cabbage production

The effect of β-CD–urea controlled-release fertilizer on Chinese cabbage (*Brassica campestris* L. ssp. *Pekinensis*) output and nitrate content has been investigated. β-CD was included at a molar ratio of 3:1 β-CD:urea, and the complex was compared with normal urea fertilizer containing 3, 6, or 9 g of urea. Cabbages were planted with experimental fertilizer for 2 months from July to September. The results of the field experiments showed that at the same urea level, cabbage output was 21.42–30.76% in β-CD–urea inclusion complex groups than normal urea groups and the nitrate content in leaves was 9.74–29.47% lower than in the normal urea groups. These results confirmed that β-CD–urea inclusion complexes can improve the output and quality of Chinese cabbage production [9].

4.2.3 Applications in hulless barley seed germination

The effects of β-CD–urea controlled-release fertilizer on seed germination and the seedling physiological index of hulless barley (*Hordeum vulgare* L. var. *nudum* Hook. f.) were investigated by pot experiments. The β-CD–urea controlled-release fertilizer was prepared at a weight ratio of 1:1 β-CD to urea, and 20 hulless barley seeds were planted per pot and treated with fertilizer containing 0, 0.67, 1.33, 2.00, and 2.67 g/kg urea. The germination rate after 7 days and the emergence rate at 15 days were 21.6% and 25.0% higher in the β-CD-urea (1.33 kg/g) group than in the equivalent pure urea group. Similarly, the seedling height was increased by 8.20 cm compared with the equivalent pure urea group. Additionally, the relative conductivity of leaves was reduced by 33.23%, the total chlorophyll content was increased by

0.284 mg/g, and the soluble protein content was increased by 6.94 mg/g, compared with the pure urea group [10]. These results showed that at the same urea level, the β-CD–urea complex was superior for seed germination, emergence rate, and seedling height. β-CD–urea controlled-release fertilizers appear to shield the toxic effects of urea, avoid urea-induced inhibition of seed germination and seedling growth, and meet the nutrient requirements for hulless barley growth.

4.2.4 *Improving the effectiveness of monoammonium phosphate in soil*

Phosphorus fertilizers can significantly increase crop yields [11–13], and monoammonium phosphate (MAP) is an important water-soluble compound that is widely used in agricultural soil [14]. However, phosphorus can easily leach from the top layer of soil after rainfall [15], and steeping causes phosphorus losses resulting from surface water eutrophication. Furthermore, water-soluble phosphate can readily react with metal cations in soil such as Ca^{2+}, Mg^{2+}, and Al^{3+}, and it is easily adsorbed onto clay minerals so that its migration and effectiveness decline with time [16].

β-CDs have been used to prepare a β-CD–MAP inclusion complex that releases phosphate more slowly. MAP aqueous solution was mixed with β-CD at a molar ratio ranging from 0:1 to 2:1, incubated at 40°C with stirring at 150 rpm for 24 h in a shaker, cooled to room temperature (25°C), and freeze-dried. The inclusion complex has an absorption maximum at 193 nm, and characterization showed that a molar ratio of β-CD to MAP of 1:1 and an inclusion time of 10 h were appropriate [4].

Analysis of adsorption and desorption revealed an adsorption capacity constant (K_f) of 3.293 for soil containing the MAP and β-CD complex, which was lower than the K_f value of 11.65 with MAP alone.

The sorption intensity constant n value (0.8050) of MAP with β-CD was higher than the n value (0.4934) of MAP without β-CD, indicating that the presence of β-CD plays an important role in adsorption of MAP on the soil surface. This is presumably because the

MAP molecule is occluded in the β-CD cavity to form the β-CD–MAP inclusion compound.

The total percentage of MAP desorbed (η) with different initial MAP concentrations (1500, 2000, and 2500 mg/L) in the presence of β-CD was 9.3%, 15.7%, and 18.6%, respectively, which was higher than in the absence of β-CD. This result indicates that the β-CD–MAP inclusion compound can be desorbed easily, possibly due to decreased chemical adsorption. Thus, the β-CD–MAP inclusion compound can prevent the reaction of MAP with metal cations in soil, thereby increasing the rate of migration of MAP in soil [4].

4.2.5 *Stabilizing urease inhibitors using CDs*

N-butyl-phosphorotriamide (NBPT) is widely used as a urease inhibitor in fertilizers. Urease inhibitors promote the assimilation of nitrogen (urea) by plants, which is beneficial for plant growth. However, current commercially available formulations are complex and do not avoid the severe decrease in activity resulting from the low stability of the bioactive compound under acidic conditions. Based on its structure, NPBT was thought to be able to interact with both polar (via its phosphoramide moiety) and hydrophobic (via its alkyl chain) additives. A panel of natural polysaccharides including starch, β-(1,3)-glucans, carrageenans, alginates, and cyclodextrins was prepared to study the ability of NBPT to protect against hydrolysis under acidic conditions [17]. The results showed that the best additive was α-CD. Solutions of α-CD in water (10 mM) and NBPT in methanol (10 mM) were prepared and different amounts of the two solutions were mixed together in test tubes (t ratios of 0:10 to 10:0) followed by freeze-drying. The association constant was determined by ^{1}H NMR, and the inhibitory activity of the inclusion compound determined by spectrophotometry. The results showed that NBPT interacted most favorably with the small α-CD containing six glucosyl residues. Furthermore, the optimal molar ratio of α-CD:NBPT was experimentally determined to be 1:1. Moderate inhibition activity (1.5 μM) was observed, and at least 70% of NBPT remained stable after 4 days at pH 5.5, while 60% of NBPT was hydrolyzed under the same conditions in the absence of α-CD. This corresponds to a 75% increase in the stability of NBPT in the inclusion complex, which is a stable, highly water-soluble solid. Application to acidic soil showed that the inclusion complex elicited a similar inhibition efficiency to commercially available liquid NBPT [17].

4.2.6 *Applications in Fe-nanosponge complex fertilizers*

Among micronutrients, iron is an essential nutrient for plant growth and development, since it participates in vital processes such as respiration and photosynthesis [18]. Increasing the amount of crop-available Fe has long been carried out by iron-containing fertilizer application to soil and irrigation water. Most fertilizers of this type are synthetic Fe chelates, which are very effective but expensive, and their persistence in soil and groundwater is a cause for environmental concern [19]. A new iron fertilizer was developed using a β-CD-based nanosponge (NS) complex (Fe-NS) [20]. CD-based NSs are biocompatible nanoporous nanoparticles that are synthesized by carbonyl or dicarboxylate hyper cross-linking of CDs [21]. β-CDs are cross-linked with pyromellitic anhydride (1:5 molar ratio) at 90°C for 2 h using dimethyl sulfoxide (DMSO) as a solvent and triethylamine as a catalyst. The compact and dense final (β-CD-based NSs have a tendency to form a gel in polar solvents and can be broken, purified with acetone, and milled to a fine powder.

Fe-NS was prepared by stirring 18.8 g of NS in 100 mL of 9.4% $FeSO_4 \cdot 7H_2O$ for 24 h. The solution was acidified with HNO_3 (pH 3) to prevent precipitation of Fe hydroxides. NSs had no significant effect on the growth of sweet corn in $FeSO_4$ nutrient solution, and in fertilization tests lasting 14 days, the dry weight of Fe-NS-treated plants (shoot = 1.01 g and root = 0.48 g) was higher than that of plants treated with $FeSO_4$ or Fe-DTPA (diethylenetriaminepentaacetic acid) (shoot = 0.56 g and 0.52 g, respectively). The chlorophyll content SPAD index value of Fe-NS-treated plants (20.69) was also higher than what was observed with the other treatments. These results indicate that Fe-NS is more effective for promoting sweet corn growth.

When Fe fertilizers were applied to tomato plants for 7 or 14 days, no significant differences were observed among treatments. However, the shoot dry weight (1.88 g) of Fe-NS-treated plants was higher than that of Fe-DTPA-treated plants, and the Fe contents in leaves of $FeSO_4$ and Fe-NS plants (182 and 172 ppm, respectively) were higher than those of Fe-DTPA plants (120 ppm).

When Fe-NS was supplied in the nutrient solution, this had a positive effect on regreening, dry matter, and Fe content in both sweetcorn and tomato plants. Moreover, Fe-NS ensured good plant growth, probably due to the gradual release of nutrients and microelements according to plant requirements [20].

4.3 Applications in soil amendment

4.3.1 *Amending the formation of water-stable aggregates in silty coastal soil*

Soil structure stability is a very important parameter among the soil physical and chemical properties, since it influences not only soil moisture and nutrient availability but also gas exchange, heat balance, soil microbial activity, and extension of the plant root system. The amount of water-stable aggregates can reflect the stability of the soil structure [22]. Silty coastal soil is an important land resource that is easily lost following rain due to the lack of inorganic colloids and low soil organic matter. Soil amendment is applied to improve the soil structure.

Soil structure amendment processes include both natural and synthetic approaches, the former including humic acid, polysaccharide, cellulose, and lignin, and the latter including polyvinyl alcohol, polyacrylamide, asphalt emulsion, and polyacrylonitrile. These amendments are characterized by high stability and low toxicity, which effectively improves soil structure and water storage capacity, and increases the soil temperature and corrosion resistance.

β-CDs have been added as a soil amendment to enhance the formation of water-stable aggregates and nutrient supply capacity in silty soil, and the results compared with the addition of the more conventional polyacrylamide and humic acid. β-CDs, polyacrylamide, or humic acid were added to silty coastal soil at a dry soil weight of 0.05%, 0.10%, 0.20%, and 0.50% (a control group without soil amendment was included). Two rounds of fertilizer and amendment were performed. The nutrients NH_4-N, NO_3-N, PO_4-P, and K were added to soil at 25, 25, 30, and 30 mg/kg on a dry soil weight basis, and the formation of ammonium chloride, sodium nitrate, calcium dihydrogen phosphate, and potassium chloride was monitored. Amendment and fertilizer were mixed with 3.5 kg of soil in a plastic pot, and the moisture level was maintained at 75% by adding deionized water over the 2-month incubation.

The results showed that the addition of amendment at levels greater than 0.10% decreased the soil density significantly compared with the control group. The order of additions had no significant effect on the soil density or the composition of water-stable aggregates. β-CDs, polyacrylamide, and humic acid all improved the formation of water-stable aggregates in soil, in the order polyacrylamide $>\beta$-CD $>$ humic acid (largest to smallest effect). The amount of water-stable aggregates with size >0.25 mm increased with increasing addition of modifiers. The optimal dosage of β-CDs and polyacrylamide for modifying soil structure was $\sim 0.20\%$. Application of β-CDs and poly-acrylamide altered the supply capacity of nutrients in soil. Fertilizing followed by addition of modifiers weakened the supply capacity of nutrients compared with controls without any amendments. However, addition of modifiers followed by fertilizing improved the supply capacity of nutrients compared with the control group [23].

4.3.2 *Decontamination of sewage sludge using hydroxypropyl-β-CD*

Urban wastewater is treated in wastewater treatment plants to avoid the spread of disease, remove organic matter and pollutants, and preserve the quality of surface water. However, more and more sewage sludge is continually generated that requires disposal. A useful and interesting option is the production of compost and the direct application of stabilized sludge to agricultural land, due to the potentially positive effects on soil fertility, modification of foil structure, and addition of organic matter and nutrients including N, P, and K. However, the important step is to eliminate the potential hazards present in sewage sludge, including persistent toxic organic compounds. Hydroxypropyl-β-CD (HPBCD) exhibits a low tendency to adsorb onto soil particles [24] and is well tolerated in humans [25]. This compound can improve water solubility and is toxicologically benign.

4.3.3 *Enhancing the biodegradation of diuron using hydroxypropyl-β-CD*

The phenylurea herbicide N-(3,4-dichlorophenyl)-N,N-dimethylurea (diuron) is widely used in a broad range of applications, and it is a biologically active toxic compound present in soil, water, and sediments. Diuron is strongly adsorbed by soil organic matter particles and slowly degraded in the environment due to its reduced bioavailability. The crucial step in increasing the bioavailability of diuron is enhancing its solubility.

HPBCD was added to soil containing diuron and incubated with a bacterial consortium comprising the diuron-degrading *Arthrobacter sulfonivorans* (*Arthrobacter* sp. N2) and the linuron-degrading *Variovorax soli* (*Variovorax* sp. SRS16). After incubation for 120 days, an almost complete (98.67%) biodegradation of diuron was achieved, whereas only 45% biodegradation was observed in the absence of the CD solution, compared with incomplete mineralization based on single or consortium bacterial degradation. These results indicate that HPBCD acts as a bioavailability enhancer, even at very low concentrations (tenfold higher than the soil diuron concentration), by accelerating diuron desorption from the soil particle surface into the soil solution, and improving the accessibility of the herbicide to microorganisms [26].

4.3.4 *Enhancing the removal of naphthalene and phenanthrene from soil*

Polycyclic aromatic hydrocarbons (PAHs) are strongly adsorbed into soils or sediments [27], and this is receiving increasing attention because of their highly carcinogenic nature and their continuous release into the environment through human activities [28]. Various physical, chemical, biological, and combined technologies have been attempted to remediate PAH-contaminated soils [29]. CDs have been proposed as an alternative agent to enhance the water solubility of hydrophobic compounds [30].

Naphthalene and phenanthrene were used as model PAHs to investigate the adsorption and desorption of in soil and the influence of CDs in this process. The solubility of naphthalene and phenanthrene was enhanced twentyfold and ninetyfold, respectively, in 50 g L^{-1} of HPBCD, which was more effective for solubilizing both compounds than β-CD at the same concentration. When HPBCD was used as a flushing agent, 80% of naphthalene and 64% of phenanthrene was recovered from soil. For both compounds, the slowest desorption rate was found for soil containing the highest organic matter content. β-CD sorption in soils was relatively low and dependent on soil type, and high soil organic matter favors the retention of both CDs and pollutants, which slows the extraction rate [24].

4.4 Applications in animal feeds

4.4.1 *Hypolipidemic effects of β-CD in a rat diet*

The effect of β-CD on plasma cholesterol and triglycerides when included in the diet were investigated in male genetically hypercholesterolemic Rico rats and male Syrian hamsters [31]. The experimental diets contained 10, 50, 100, or 200 g/kg β-CD. Thirty hamsters and 18 Rico rats that were 12 weeks old at the beginning of the experiment received food and water throughout the 8 weeks of the study. In both Rico rats and hamsters, plasma cholesterol and triglycerides decreased linearly with an increasing dose of β-CD. In these two species, 20% β-CD in the diet lowered cholesterolemia by 35% and triglyceridemia by 70% compared with controls [31].

The observed decrease in cholesterol absorption was probably linked to the encapsulating capacity of β-CD for cholesterol and/or to the disruption of intestinal bile salt mixed micelles following bile acid complexation. Complexation with β-CD has been observed *in vitro* for various bile acids in aqueous medium [32].

The strong cholesterol–β-CD interaction occurs in intestinal compartments, and such interactions could be induced either by cholesterol encapsulation, which would prevent its microbial degradation, or by a change in the composition of the β-CD-fermenting microflora, which may eliminate the bacteria responsible.

These results suggest that β-CD does not alter the microbial degradation of bile acids but rather stimulates their synthesis and increases their pool size. β-CD prevents the intestinal absorption of lithocholic acid and washes this cytotoxic bile acid from the colon. The hypolipidemic effect of β-CD was shown to be due to the stimulation of bile acid biosynthesis and a decrease in cholesterol absorption, with the former mechanism more important for the elimination of cholesterol from plasma [33].

4.4.2 *Dietary β-CD reduces cholesterol levels in pigs*

Hypocholesterolemic and triacylglyceride lipid-reducing effects of β-CD have been reported in hamsters [31], swine [34], and mice [35]. These effects are due to the fact that the hydrophobic base of the inner core of β-CD maintains a high affinity for cholesterol *in vivo*, thereby inhibiting the absorption of lipids and stimulating the binding of bile acids or neutral sterols in the gut and increasing steroid excretion through the feces [34]. In one study, β-CD was added in the diet of swine to determine the hypocholesterolemic effect [36]. A total of 120 pigs were assigned to four dietary groups receiving 0% (control), 5.0%, 7.0%, or 10.0% β-CD in the diet for 30 days. The results showed an increase in body weight from 85 kg to 110 kg. The diets were rich in cholesterol from 5.31 to 10.58 mg g^{-1} for four of the diet groups, and all pigs had *ad libitum* access to commercial grower and finisher phase diets and water. After slaughter, pigs were dissected into individual muscle and fat samples (back fat, loin, belly, and ham).

There results revealed no significant difference in feed intake (3.01–3.25 kg day^{-1}), daily gain (798–807 g day^{-1}), and feed conversion ratio (3.76–4.03 g g^{-1}), indicating that dietary β-CD (at 5%, 7%, or 10%) did not affect the performance of finishing pigs. Total plasma lipids were reduced by 16.0%, 18.7%, and 20.8% in pigs fed diets supplemented with 5%, 7%, and 10% β-CD, respectively. A 55.3% reduction in plasma triglyceride concentration was observed in pigs fed diets supplemented with 10% β-CD. Total plasma cholesterol levels in control, 5%, 7%, and 10% β-CD groups were 96.4, 71.3, 69.2, and 66.2 mg d L^{-1}, respectively. The plasma cholesterol level decreased with increasing supplementation of β-CD. This indicated that β-CD is very effective in reducing plasma cholesterol in finishing pigs supplemented with cholesterol-enriched diets.

β-CD has a remarkable ability to form inclusion complexes with various natural and synthetic molecules; in particular, it is capable of encapsulating and solubilizing cholesterol and bile acids to varying degrees *in vitro*. Supplementing the finishing diet with 5%, 7%, or 10% β-CD decreased the cholesterol content in back fat by 24.1%, 28.0%, and 31.2%, respectively, whereas cholesterol levels in the loin were reduced by 29.2%, 29.9%, and 32.9%, respectively. The cholesterol concentration in the bellies of pigs supplemented with 10% β-CD was decreased by 23.1% compared to that of controls.

In conclusion, the cholesterol concentrations in pork lean meat and fat was markedly reduced by supplementing swine diets with β-CD, which may provide the swine industry with a method for developing novel, low-cholesterol pork products [36].

4.4.3 *Beneficial effects of supplemental γ-CD on nutrient digestibility and microbial populations in dogs*

The effects of supplementation of γ-CDs on nutrient digestibility, microbial populations, and fecal characteristics were investigated in dogs [37]. The bacterial population in the gastrointestinal tract of mammals constitutes a metabolically active community that acts as a significant barrier to infection by exogenous pathogenic microorganisms. Furthermore, the gut bacterial population can have both a beneficial or detrimental effect of the host [38]. Beneficial microbial populations are important in resisting pathogenic species and producing short-chain fatty acids. Both γ-CD and pullulan are incompletely digestible nonstructural carbohydrates and may be beneficial to gut microbial populations.

In one study, γ-CD (2 or 4 g per day) was fed to dogs via gelatin capsules in combination with a commercial dog diet, supplemented with and without pullulan. Dogs averaged 3.7 years of age, had an initial average body weight of 28.9 kg, and were fed a 200 g diet twice daily for 14 days. The results showed that supplementation with γ-CD and pullulan did not influence the ileal or total tract nutrient digestibility. For dogs receiving 2 or 4 g γ-CD per day, fat digestibility between the ileum and feces was increased by approximately 2% and 1.5%, respectively, and carbohydrate digestibility was increased by approximately 2% and 3.5%, respectively. Dogs receiving 2 g of γ-CD per day had higher concentrations of bifidobacteria and lactobacilli and lower concentrations of fecal *Clostridium perfringens*, compared with both 0 g and 4 g γ-CD treatment groups.

Both γ-CD and pullulan may improve gut health by altering microbial populations in a positive direction, which may be particularly beneficial to geriatric dogs, puppies, or dogs under stress that have compromised gut microbial populations [37].

4.4.4 *Reducing the cholesterol content of egg yolk by β-CD supplementation in the hen diet*

Hen eggs are a particularly nutritious natural food source, but their intake should be limited because they contain ~200–250 mg of cholesterol [39]. The composition of the feed of laying hens has been explored to lower the cholesterol content of eggs [40, 41]. The effect of feeds containing β-CD has also been investigated [42].

In one study, five dietary groups of 20 hens were fed a diet containing 0%, 2%, 4%, 6%, or 8% β-CD that mainly consisted of corn and soybean meal. Soybean oil was provided at 15 g per kg in the experimental diet. Each group of layers was housed in a laying cage and maintained for 10 weeks until they reached 48 weeks.

The results showed that the group receiving 8% β-CD consumed less food than the other groups. The egg production rate, egg weight, and feed intake were constant in all groups except the 8% β-CD group. Because the feed intake was lower in the 8% β-CD group, their feed requirements per egg produced were lower. In animals fed large quantities of β-CD, a metabolic disorder probably occurs in the small intestine such that palatability of the diet is decreased.

The major finding was the observation that egg cholesterol was markedly and progressively reduced as the β-CD content of the diet was increased. The yolk cholesterol content was decreased from 14.16 mg/g yolk (0% β-CD group) to 9.92 mg/g yolk (8% β-CD group). This may be related to the fact that β-CD has a molecular structure that accommodates the cholesterol molecule within the hydrophobic cavity of the doughnut, and the tight insoluble complex formed prevents cholesterol absorption in the gut [43]. There was no statistical significance in the Haugh unit, egg shell thickness, or egg yolk color between control and β-CD-treated hens.

4.5 Applications in pesticides

CD contains a unique hydrophobic cavity that can form inclusion complexes with various guest molecules. For instance, pesticides can be integrated into CD-inclusion complexes to improve their solubility, irritating effects, chemical stability, drug release rate, and bioavailability. Therefore, CD plays an important role in pesticide formulations. This chapter describes the effects of CD and its derivatives on the properties of pesticides, such as stabilization, solubilization and light-catalytic degradation. This chapter also introduces the application of CD for the chiral separation of pesticides and for the detection of pesticide residues.

4.5.1 *Stabilization*

Chemical pesticides are mostly composed of organic compounds. However, these substances are unstable; as such, they hydrolyze and ionize in aqueous solutions, decompose in the presence of light, react with coexisting substances, decompose during processes and volatilize. Pesticides can be integrated with CD via van der Waals force and hydrogen bond, and their reactive sites become wrapped in the hydrophobic cavity of CD. Thus, their stabilization and utilization are improved because of the reduced contact with external environmental factors, such as light, heat, and humidity.

CD is mainly used as a stabilizer of herbicides, pesticides, and plant growth agents to maintain long-term efficiency. Pesticides with low toxicity and high efficiency are usually synthetic insecticides accounting for approximately 20% of the pesticide market, but many insecticides become invalid when they are exposed to light. However, CD-inclusion stabilization can be applied to store and improve the efficiency of insecticides. Kamiya *et al.* [44] found that the formation of β-CD-inclusion complexes with organophosphorus insecticides, such as parathion, and their oxidation products, such as paraoxon, can stabilize parathion; once stabilised, parathion prevents the generation of toxic paraoxon and reduces environmental pollution.

4.5.2 *Solubilization of hydrophobic pesticides*

Mass-produced pesticides in China are classified as hydrophobic organic substances. In aqueous solutions, these pesticides are ineffectively dissolved, and pesticides are mostly used in an emulsion form. Emulsion preparation consumes large amounts of organic solvents, and this mechanism contradicts current development trends of pesticides. Water preparation rather than emulsion is an effective method to address concerns on pesticide safety and environmental pollution. With a hydrophilic outer wall and a hydrophobic inner cavity, CD can be used as a solvent to solubilize hydrophobic pesticides.

The solubilization of CD in pesticides is mainly caused by the formation of host–guest inclusion complexes, such as CD-pesticides, which increase the apparent solubility of pesticides. CD–pesticide inclusion complexes exhibit higher water solubility than individual fungicides do. Li *et al.* [45] reported that the solubility of trimethoprim (TMP)–β-CD-inclusion compound formed by TMP and CD increases by 2.2 times compared with TMP alone. Lezcano *et al.* [46, 47] selected eight kinds of low-water-solubility bactericidal agents, used a CD inclusion to enhance water solubility and investigated the feasibility of preparing solid-phase CD-bactericidal agents. They reported that the water solubility of the inclusion compounds formed by three hydrophobic benzimidazole bactericidal agents and β-CD is significantly improved.

The solubilization effect of β-CD on pesticides is mainly determined by the size, hydrophobicity, and matching degree of pesticides and the cavity of β-CD; for instance, high matching corresponds to a remarkable solubilization effect. The water solubility of β-CD is low (18.5 g/L, 20°C) because of the formation of an intramolecular hydrogen bond between C2 and C3; as a consequence, β-CD fails to increase the water solubility of insoluble compounds. The solubilization effect of β-CD on pesticides can be significantly increased by modifying the outer surface of the CD cavity through derivatization and other processes, including hydroxyl etherification, esterification, and oxidation. Therefore, modified CD can be used to enhance the water solubility of hydrophobic pesticides.

Diclofop-methyl (DM) is an herbicide with a broad spectrum but with poor water solubility (3 mg/L, 20°C). However, this herbicide can be transported slowly in plants *in vivo*; as such, this limitation greatly reduces its weeding performance. β-CD and its derivatives (CM-β-CD, 2-HP-β-CD and methyl-β-CD) can significantly increase the water solubility of DM; among these derivatives, methyl-β-CD elicits a good linear solubilization effect on DM [48].

Luo *et al.* [49] reported that hydroxyl radical β-CD (HPCD) can improve the water solubility of three hydrophobic organic pesticides, namely, methyl parathion, carbofuran, and pentachlorophenol; they also revealed that the solubilization effect of HPCD on pesticide is significantly higher than that of β-CD. Liu *et al.* [50] further reported that methylated β-CD (MCD) increases the water solubility of pesticides; in 60 g/L MCD solution at 25°C, the water solubilities of methyl parathion and carbofuran are increased by 63.97 and 92.3 times, respectively. Hou *et al.* [51] reported that modified CD can significantly increase the water solubility of cyhalothrin.

4.5.3 *Photocatalytic degradation*

Currently used pesticides are mostly composed of hydrophobic organic substances that can be easily adsorbed by soil colloid; as a result, these pesticides cannot be easily transferred, degraded, and accumulated in soil. By comparison, CD derivatives are water soluble, nontoxic, and biodegradable; they cannot be absorbed by the soil. Thus, CD does not accumulate in soil and other media, and the occurrence of secondary pollution is prevented. CD-inclusion compounds are formed by the hydrophobic cavity of CD and pesticides; once formed, these compounds can desorb the hydrophobic pesticides adsorbed by soil organic matter and reduce pesticide residue concentrations [52].

Methyl parathion and pentachlorophenol are representatives of highly toxic pesticides among the organic phosphorus and chlorine and are widely used in China. Zeng *et al.* [53] reported the photocatalytic degradation of MCD and HPCD on methyl parathion. Tang *et al.* [54] reported that the photolysis effect of HPCD on methyl parathion and pentachlorophenol is more significant than that of β-CD. Liu *et al.* found that MCD or HPCD increases the photolytic rate by 4.87 and 6.85 times, respectively.

The photo degradation of β-CD on pesticides is also influenced by other factors. Kamiya *et al.* [55] demonstrated that the photodegradation of organic phosphorus pesticides in humic water is affected by the production of free radicals and the size of the CD cavity. Ishiwata *et al.* [56] showed that CD can improve the inclusion-catalytic activity of organophosphorus pesticides in a neutral medium and inhibit the degradation of pesticides in an alkaline medium. This phenomenon occurs possibly because the electrostatic repulsion between ionized hydroxyl groups in an alkaline medium enlarges the CD aperture. As a consequence, the inclusion of the pesticide becomes deep. Thus, the degradation of pesticides is suppressed. In addition, the catalytic effect of CD is greatly influenced by the properties of its substituent group, aromatic ring, heterocyclic ring, and cavity size. The relationship of the structure of CD inclusions with the catalytic degradation of pesticides is being investigated further [57].

4.5.4 *Separation of chiral pesticides*

With advancements in three-dimensional chemistry, the relationship between the molecular structure and biological activity of compounds has been extensively investigated. Approximately 25% of pesticides contain chiral centers, and most of these centers are composed of carbon or phosphorus atoms in organic phosphorus pesticides. Chiral pesticides are usually racemic, and one of the isoforms may be harmful to nontarget materials. The biodegradation and metabolism of chiral pesticides can also exhibit stereoisomerism [58]. Single chiral pesticides may exhibit high efficacy at low dosage; these pesticides may also release low amounts of waste, promote crop and ecological environment security, entail low costs, and display strong market competitiveness. Therefore, pesticides should be chirally separated.

The physical and chemical properties of chiral enantiomers are very similar; researchers experience difficulty in separating them because they can be distinguished in a chiral environment [59]. CD is a truncated cone-like macrocyclic molecule with a three-dimensional chiral hydrophobic cavity; it can provide the enantiomers with a combined space [60]. CD derivatives can form inclusion complexes with chiral molecules, as confirmed through X-ray and ^{1}H NMR. Thermodynamic data have revealed that the change in the total entropy of the enantiomers during separation is considerably high because of the inclusion complex. The structure of CD is modified on the basis of different pesticide molecules to obtain CD derivatives. In this manner, chiral molecules can be separated. Thus, they become suitable components of the chiral part of CD derivatives. Their separation ability is also enhanced.

4.5.4.1 *CD as a chiral selector for capillary electrophoresis*

In chiral separation, CD is commonly used as a chiral selector; β-CD and its derivatives are the most widely used chiral additives for chiral separation in capillary electrophoresis [61]. Chemically modified β-CD derivatives have been extensively investigated to enhance the water solubility and chiral resolution ability of CD derivatives by introducing specific structures of side chains or groups [62].

Capillary electrophoresis (CE) is also known as high-performance capillary electrophoresis (HPCE). In CE, CD and its derivatives are used as chiral selectors. This method is characterized by high efficiency and fast and simple operation. CD-modified capillary zone electrophoresis has been successfully applied to separate isomers in pesticides, and the separation mechanism is based on different forces between separated substances and CD-inclusion compounds. The isolation efficiency is affected by buffer pH, CD type, and concentration [63].

A dextroisomer with a weed activity is separated from the racemate enantiomers of 2,4-dichlorprop through capillary zone electrophoresis modified by α-, β-, and γ-CD [64]. Zerbinati *et al.* [65] isolated four herbicide isomers through capillary zone electrophoresis modified by CD as a solution. Zhou [66] successfully separated four kinds of zeolite antibacterial agents in 15 min by using HP-β-CD as a chiral selector. Shi *et al.* [67] successfully separated four kinds of pesticide intermediate enantiomers with optical activity, namely, 2-phenoxy propionic acid, *cis*-cyhalothric acid, 1-phenyl-2-(*p*-tolyl) ethylamine, and 1-phenyl-2-(methoxy phenyl) ethylamine by using β-CD polymer as a chiral selector.

4.5.4.2 *CD as a chromatographic stationary phase for the separation of pesticide optical isomers*

In pesticides, some optical isomers exhibit good biological activity. For instance, chrysanthemic acid pesticides, developed in the 1980s, exhibit high efficiency and low toxicity; these pesticides are also widely used in agriculture and as health insecticides. The biological activities of various stereoisomers significantly differ. For example, deltamethrin is composed of three chiral centres and eight isomers, and the insecticidal activity of the highest isomer (3R, 1R, S) is greater than that of the worst isomer (3S, 1S, R) by 70 times. The unique cavity of CD can contain many organic compounds; the molecule also comprises numerous chiral centres. β-CD consists of as many as 35 chiral carbons. CD and its derivatives are widely used in chromatographic chiral separation not only for ordinary compounds, such as nonpolar hydrocarbons, highly polar polyols, and amino acids, but also for optically active compounds. In the chiral separation of a special stationary phase, CD and its derivatives yield good separation efficiency and promote an efficient baseline separation of optical isomers from 20 kinds of pesticides and intermediates of warfarin and *cis–trans*-chrysanthemic acid methyl-2-chloride [68]. Shi *et al.* [69, 70] used four kinds of synthesized acyl CD derivatives as a stationary phase of capillary gas chromatography and achieved good separation effects of the enantiomers from six kinds of pyrethroic acid methyl ester.

4.5.5 *Application of capillary electrophoresis modified by CD in the detection of pesticide residues*

CE is characterised by high sensitivity, high separation rate, high analysis speed, low sample consumption, high automation degree, and low costs [71]. HPCE is applied for the separation analysis of pesticides, especially various herbicides; HPCE is also employed to prepare single-species pesticides and to determine active components in compound pesticides [72].

CD-modified CE can be applied to detect pesticide residues [73]. Using micellar electrokinetic chromatography method, Schmitt *et al.* [74] separated and detected crufomate, isofenphos, chlorine, phosmet, benzene, phosphorus, and malathion in boric acid buffer solution with SDS and DM-β-CD. Hsieh *et al.* [75] also utilised CE to separate seven chlorinated phenoxy acid herbicides and enantiomeric isomers in a buffer with α- and β-CD for 7 min.

4.6 Conclusion

The relationship between biological activity and compound structure has been extensively investigated. Enantiomers with high biological activities have been used to develop modern pesticides. Considering the unique physical, chemical, and biological characteristics of CD, researchers should further explore the chiral separation mechanisms of CD and pesticides. We should also analyze the optical purity of chiral pesticides by using simple, convenient, and reliable determination methods.

References

1. Zou CJ, Liao WJ, Zhang L, Chen HM. (2011). Study on acidizing effect of β-cyclodextrin-PBTCA inclusion compound with sandstone. *Journal of Petroleum Science and Engineering*, 77(2), pp. 219–225.
2. Gao YA, Li ZH, Du JM, Han BX, Li GZ, Hou WG, Shen D, Zheng LQ, Zhang GY. (2005). Preparation and characterization of inclusion complexes of β-cyclodextrin with ionic liquid. *Chemistry: A European Journal*, 11(20), pp. 5875–5880.
3. Horiuchi Y, Abe K, Hirayama F, Uekama K. (1991). Release control of theophylline by beta-cyclodextrin derivatives: hybridizing effect of hydrophilic, hydrophobic and ionizable beta-cyclodextrin complexes. *Journal of Controlled Release*, 15(2), pp. 177–183.
4. Liao W, Tang D, Huang X, Wang H, Dang X. (2015). Self-improvement value of monoammonium phosphate by complexation effect of β-cyclodextrin in soil. *Industrial & Engineering Chemistry Research*, 54(38), pp. 9263–9269.
5. Liao W, Tang D, Cheng X. (2014). Preparation of β-cyclodextrin-urea controlled-release fertilizers and its application performance. *Journal of Anhui Agricultural Science*, 42(13), pp. 3823–3824, 3827.
6. Yang Y, Zhang G, Shuang S, Dong Z, Pan J. (2003). Development on inclusion interaction and molecular recognition of cyclodextrin. *Chinese Journal of Spectroscopy Laboratory*, 20(20), pp. 169–180.
7. Li GH, Zhao LP, Zhang SX, Hosen Y, Yagi K. (2011). Recovery and leaching of N-15-labeled coated urea in a lysimeter system in the North China Plain. *Pedosphere*, 21(6), pp. 763–772.
8. Soares JR, Cantarella H, Menegale MLD. (2012). Ammonia volatilization losses from surface-applied urea with urease and nitrification inhibitors. *Soil Biology & Biochemistry*, 52, pp. 82–89.
9. Cheng X. (2015). Effect of β-cyclodextrin-urea controlled-release fertilizer on Chinese cabbage output and nitrate content. *Hubei Agricultural Sciences*, 54(22), pp. 5565–5568.
10. Liao W. (2016). Effects of β-cyclodextrin controlled release fertilizer on seed germination and seedling physiological index of hulless barley. *Guizhou Agricultural Sciences*, 44(5), pp. 28–31.
11. Li CY, Li C, Zhang RQ, Liang W. (2013). Effect of phosphorus on the characteristics of starch in winter wheat. *Starch-Stärke*, 65(9–10), pp. 801–807.
12. Noda T, Tsuda S, Mori M, Suzuki T, Takigawa S, Matsuura-Endo C, Yamauchi H, Sarker MZI. (2012). Effects of annual fluctuation of environmental factors on starch properties in potato tuber development. *Starch-Stärke*, 64(3), pp. 229–236.
13. Waterschoot J, Gomand SV, Fierens E, Delcour JA. (2015). Production, structure, physicochemical and functional properties of maize, cassava, wheat, potato and rice starches. *Starch-Stärke*, 67(1–2), pp. 14–29.
14. Ma YB, Li JM, Li XY, Tang X, Liang YC, Huang SM, Wang BR, Liu H, Yang XY. (2009). Phosphorus accumulation and depletion in soils in wheat-maize

cropping systems: modeling and validation. *Field Crops Research*, 110(3), pp. 207–212.

15. Matula J. (2009). Possible phosphorus losses from the top layer of agricultural soils by rainfall simulations in relation to multi-nutrient soil tests. *Plant Soil and Environment*, 55(12), pp. 511–518.

16. Schroder JJ, Smit AL, Cordell D, Rosemarin A. (2011). Improved phosphorus use efficiency in agriculture: a key requirement for its sustainable use. *Chemosphere*, 84(6), pp. 822–831.

17. Pro D, Huguet S, Arkoun M, Nugier-Chauvin C, Garcia-Mina JM, Ourry A, Wolbert D, Yvin JC, Ferrieres V. (2014). From algal polysaccharides to cyclodextrins to stabilize a urease inhibitor. *Carbohydrate Polymers*, 112, pp. 145–151.

18. Martinez-Cuenca MR, Forner-Giner MA, Iglesias DJ, Primo-Millo E, Legaz F. (2013). Strategy I responses to Fe-deficiency of two citrus rootstocks differing in their tolerance to iron chlorosis. *Scientia Horticulturae*, 153, pp. 56–63.

19. Sanchez-Alcala I, Bellon F, del Campillo MC, Barron V, Torrent J. (2012). Application of synthetic siderite (FeCO$_3$) to the soil is capable of alleviating iron chlorosis in olive trees. *Scientia Horticulturae*, 138, pp. 17–23.

20. Vercelli M, Gaino W, Contartese V, Gallo L, Di Carlo S, Tumiatti V, Larcher F, Scariot V. (2015). Preliminary studies on the effect of Fe-nanosponge complex in horticulture. *Acta Scientiarum Polonorum-Hortorum Cultus*, 14(2), pp. 51–58.

21. Trotta F, Tumiatti W. (2003). Cross-linked polymers based on cyclodextrin for removing polluting agents, WO 03/085002A1.

22. Amezketa E. (1999). Soil aggregate stability: a review. *Journal of Sustainable Agriculture*, 14(2–3), pp. 83–151.

23. Xie G, Ji S, Kong Z, Ying, J. (2015). Effects of amendments on formation of water-stable aggregates and supply capacity of nutrients in silty coastal soil. *Chinese Journal of Agriculture*, 5(1), pp. 46–50.

24. Badr T, Hanna K, de Brauer C. (2004). Enhanced solubilization and removal of naphthalene and phenanthrene by cyclodextrins from two contaminated soils. *Journal of Hazardous Materials*, 112(3), pp. 215–223.

25. Gould S, Scott RC. (2005). 2-hydroxypropyl-beta-cyclodextrin (HP-beta-CD): a toxicology review. *Food and Chemical Toxicology*, 43(10), pp. 1451–1459.

26. Villaverde J, Posada-Baquero R, Rubio-Bellido M, Laiz L, Saiz-Jimenez C, Sanchez-Trujillo MA, Mori E. (2012). Enhanced mineralization of diuron using a cyclodextrin-based bioremediation technology. *Journal of Agricultural and Food Chemistry*, 60(40), pp. 9941–9947.

27. Magee BR, Lion LW, Lemley AT. (1991). Transport of dissolved organic macromolecules and their effect on the transport of phenanthrene in porous-media. *Environmental Science & Technology*, 25(2), pp. 323–331.

28. Wagrowski DM, Hites RA. (1996). Polycyclic aromatic hydrocarbon accumulation in urban, suburban, and rural vegetation. *Environmental Science & Technology*, 31(1), pp. 279–282.

29. Fountain JC, Klimek A, Beikirch MG, Middleton TM. (1991). The use of surfactants for in situ extraction of organic pollutants from a contaminated aquifer. *Journal of Hazardous Materials*, 28(3), pp. 295–311.

30. Ko SO, Schlautman MA, Carraway ER. (1999). Partitioning of hydrophobic organic compounds to hydroxypropyl-beta-cyclodextrin: experimental studies and model predictions for surfactant-enhanced remediation applications. *Environmental Science & Technology*, 33(16), pp. 2765–2770.
31. Riottot M, Olivier P, Huet A, Caboche JJ, Parquet M, Khallou J, Lutton C. (1993). Hypolipidemic effects of β-cyclodextrin in the hamster and in the genetically hypercholesterolemic rico rat. *Lipids*, 28(3), pp. 181–188.
32. Tan XY, Lindenbaum S. (1991). Studies on complexation between β-cyclodextrin and bile-salts. *International Journal of Pharmaceutics*, 74(2–3), pp. 127–135.
33. Khallou J, Riottot M, Parquet M, Verneau C, Lutton C. (1991). Biodynamics of cholesterol and bile-acids in the lithiasic hamster. *British Journal of Nutrition*, 66(3), pp. 479–492.
34. Ferezou J, Riottot M, Serougne C, CohenSolal C, Catala I, Alquier C, Parquet M, Juste C, Lafont H, Mathe D, Corring T, Lutton C. (1997). Hypocholesterolemic action of β-cyclodextrin and its effects on cholesterol metabolism in pigs fed a cholesterol-enriched diet. *Journal of Lipid Research*, 38(1), pp. 86–100.
35. Olivier P, Verwaerde F, Hedges AR. (1991). Subchronic toxicity of orally-administered beta-cyclodextrin in rats. *Journal of the American College of Toxicology*, 10(4), pp. 407–419.
36. Park BS, Jang A. (2008). Dietary β-cyclodextrin reduces the cholesterol levels in meats and backfat of finishing pigs. *Journal of the Science of Food and Agriculture*, 88(5), pp. 813–818.
37. Spears JK, Karr-Lilienthal LK, Fahey GC. (2005). Influence of supplemental high molecular weight pullulan or γ-cyclodextrin on ileal and total tract nutrient digestibility, fecal characteristics, and microbial populations in the dog. *Archives of Animal Nutrition*, 59(4), pp. 257–270.
38. Gibson GR, Fuller R. (2000). Aspects of *in vitro* and *in vivo* research approaches directed toward identifying probiotics and prebiotics for human use. *Journal of Nutrition*, 130(2), pp. 391S–395S.
39. Weggemans RM, Zock PL, Katan MB. (2001). Dietary cholesterol from eggs increases the ratio of total cholesterol to high-density lipoprotein cholesterol in humans: a meta-analysis. *American Journal of Clinical Nutrition*, 73(5), pp. 885–891.
40. Ouyang K, Wang WJ, Xu MS, Jiang Y, Shangguan XC. (2004). Effects of different oils on the production performances and polyunsaturated fatty acids and cholesterol level of yolk in hens. *Asian-Australasian Journal of Animal Sciences*, 17(6), pp. 843–847.
41. Lien TF, Wu CP, Lu JJ. (2003). Effects of cod liver oil and chromium picolinate supplements on the serum traits, egg yolk fatty acids and cholesterol content in laying hens. *Asian-Australasian Journal of Animal Sciences*, 16(8), pp. 1177–1181.
42. Park BS, Kang HK, Jang A. (2005). Influence of feeding β-cyclodextrin to laying hens on the egg production and cholesterol content of egg yolk. *Asian-Australasian Journal of Animal Sciences*, 18(6), pp. 835–840.

43. Yen GC, Chen CJ. (2000). Effects of fractionation and the refining process of lard on cholesterol removal by β-cyclodextrin. *Journal of Food Science*, 65(4), pp. 622–624.

44. Kamiya M, Nakamura K. (1995). Cyclodextrin inclusion effects on photodegradation rates of organophosphorus pesticides. *Environment International*, 21(3), pp. 299–304.

45. Li N, Zhang YH, Wu YN, Xiong XL, Zhang YH. (2005). Inclusion complex of trimethoprim with β-cyclodextrin. *Journal of Pharmaceutical and Biomedical Analysis*, 39(3–4), pp. 824–829.

46. Lezcano M, Novo M, Al-Soufi W, Rodríguez-Núñez E, Tato JV. (2003). Complexation of several fungicides with β-cyclodextrin: determination of the association constants and isolation of the solid complexes. *Journal of Agricultural and Food Chemistry*, 51(17), pp. 5036–5040.

47. Lezcano M, Al-Soufi W, Novo M, Rodríguez-Núñez E, Tato JV. (2002). Complexation of several benzimidazole-type fungicides with α- and β-cyclodextrin. *Journal of Agricultural and Food Chemistry*, 50(1), pp. 108–112.

48. Cai XY, Liu WP, Chen SW. (2005). Environmental effects of inclusion complexation between methylated β-cyclodextrin and diclofop-methyl. *Journal of Agricultural and Food Chemistry*, 53(17), pp. 6744–6749.

49. Luo YC, Zeng QR, Liao BH, Yang RB. (2001). Effects of β-cyclodextrin and its derivatives on solubilization and toxicity of several hydrophobic organic pesticides. *Acta Scientiae Circumstantiae*, 21, pp. 101–105 (in Chinese).

50. Liu CE, Zeng QR, Duan CQ, Guo ZY, Qin PF. (2004). Preparation of MCD and its solubilization to methyl parathion and Kebaiwei. *Yunnan Environmental Science*, 23(1), pp. 6–9 (in Chinese).

51. Hou HE, Zhu FY, Hu HT. (2003). Study on the inclusion action of cyhalothrin and β-cyclodextrin in aqueous solution. *Pesticide*, 42(2), pp. 17–19.

52. Villaverde J. (2007). Time-dependent Sorption of Norflurazon in Four Different Soils: Use of β-cyclodextrin solutions for remediation of pesticide-contaminated soils. *Journal of Hazardous Materials*, 142, pp. 184–190.

53. Zeng QR, Luo YC, Liu CE, Zhou XH. (2003). Effects of two highly water-soluble β-cyclodextrins on the solubilization and photodegradation of parathion-methyl. *Chinese Journal of Pesticide Science*, 5(3), pp. 59–64 (in Chinese).

54. Tang C, Zeng QR, Liao BH, Liu CE, Luo YC. (2005). Different photodegradative specificity of β-cyclodextrin and its derivative with two pesticides. *Journal of Molecular Science*, 21(3), pp. 52–57 (in Chinese).

55. Kamiya M, Kameyama K, Ishiwata S. (2001). Effects of cyclodextrins on photodegradation of organophosphorus pesticides in humic water. *Chemosphere*, 42, pp. 251–255.

56. Ishiwata S, Kamiya M. (1999). Cyclodextrin inclusion: catalytic effects on the degradation of organophosphorus pesticides in neutral aqueous solution. *Chemosphere*, 39(10), pp. 1595–1600.

57. Ishiwata S, Kamiya M. (2000). Structural study on inclusion complexes of cyclodextrins with organophosphorus pesticides by use of rotational strength analysis method. *Chemosphere*, 41, pp. 701–704.

58. Fidalgo-Used N, Blanco-Gonzaez E, Sanz-Medel A. (2006). Evaluation of two commercial capillary columns for the enantioselective gas chromatographic separation of organophosphorus pesticides. *Talanta*, 70, pp. 1057–1063.
59. Blanco M, Valverde I. (2003). Choice of chiral selector for enantioseparation by capillary electrophoresis. *Trends in Analytical Chemistry*, 22(7–8), pp. 428–439.
60. Wang HD, Chu LY, Chen WM, Xie R, Ju XJ, Zhang J, Yang M. (2004). Development of the application of cyclodextrin in enantioseparation. *Journal of Filtration & Separation*, 14(4), pp. 1–4.
61. Horáková H, Vespalec R. (2007). Influence of substituents on selectivity and efficiency of chiral separations of anions containing single nido-7, 8-dicarbaundecaborane cluster with α-cyclodextrin. *Chromatography A*, 1143, pp. 143–152.
62. Sun QG, Liu CH, Xu JM, Wu QY, Cai YF, Yu SC. (2006). Preparation of β-cyclodextrin sulfate and its application in enantiomeric separation of ephedrine and pseudoephedrine by capillary zone electrophoresis. *Academic Journal of Second Military Medical University*, 27(4), pp. 453–454.
63. Kumar V, Penmetsa RB, Leidy DS. (1997). Enantiomeric and isomeric separation of herbicides using cyclodextrin-modified capillary zone electrophoresis. *Journal of Chromatography A*, 790, pp. 225–234.
64. Zerbinati O, Trotta F, Giovannoli C, Baggiani C, Giraudi G, Vanni A. (1998). New derivatives of cyclodextrins as chiral selectors for the capillary electrophoretic separation of dichlorprop enantiomers. *Journal of Chromatography A*, 810, pp. 193–200.
65. Zerbinati O, Trotta F, Giovannoli C. (2000). Optimization of the cyclodextrin-assisted capillary electrophoresis separation of the enantiomers of phenoxyacid herbicides. *Journal of Chromatography A*, 875, pp. 423–430.
66. Zhou S, Ouyang J, Baeyens WRG, Zhao H, Yang Y. (2006). Chiral separation of four fluoroquinolone compounds using capillary electrophoresis with hydroxypropyl-β-cyclodextrin as chiral selector. *Journal of Chromatography A*, 1130, pp. 296–301.
67. Shi XY, Liang P, Song DL, Gao XW, Fu RN. (2004). The enantioseparation of four pesticide intermediates by capillary zone electrophoresis using neutral β-cyclodextrin polymer as chiral selector. *Chinese Journal of Analytical Chemistry*, 32(11), pp. 1421–1425 (in Chinese).
68. Ling Y, Fu RN. (1997). Application of cyclodextrin derivatives chiral stationary phase to separate optical isometric compounds of pesticide in chromatography. *Pesticide Science and Administration*, 61(1), pp. 3–5 (in Chinese).
69. Shi XY, Guo HC, Qiao Z, Hou SC, Wang M. (2001). The study on the determination of the optical purity of permethrinic acid by capillary gas phase chromatography. *Chinese Journal of Pesticide Science*, 3(4), pp. 93–96 (in Chinese).
70. Shi XY, Wang M, Guo HC, Qiao Z, Ling Y, Li N, Jin WG, Zhou JM. (2002). Enantioseparation of several pyrethroic acid methyl esters on capillary gas chromatography. *Chinese Journal of Analytical Chemistry*, 30(11), pp. 1293–1297 (in Chinese).

71. Xie XM, Liao M. (2004). Quantitative determination of sulfonylurea herbicides in a paddy soil at trace level by capillary electrophoresis. *Chinese Journal of Pesticide Science*, 6(2), pp. 57–61 (in Chinese).
72. Tang JS, Xiang L. (2005). Application of capillary electrophoresis to pesticide residues determination. *Analysis and Testing Technology and Instruments*, 11(3), pp. 215–220.
73. Picó Y, Rodríguez R, Mañes J. (2003). Capillary electrophoresis for the determination of pesticide residues. *Trends in Analytical Chemistry*, 22(3), pp. 133–151.
74. Schmitt P, Garrison AW, Freitag D, Kettrup A. (1997). Application of cyclodextrin-modified micellar electrokinetic chromatography to the separations of selected neutral pesticides and their enantiomers. *Journal of Chromatography A*, 792, pp. 419–429.
75. Hsieh Y, Huang H. (1996). Analysis of chlorophenoxy acid herbicides by cyclodextrin-modified capillary electrophoresis. *Journal of Chromatography A*, 745, pp. 217–223.

5. Applications in Pharmaceuticals

Xiuting Hu* and Yaoqi Tian[†]

*School of Food Science and Technology,
Nanchang University,
Nanchang 330047, China
[†]The State Key Laboratory of Food Science
and Technology, Jiangnan University,
Wuxi 214122, China

5.1 Introduction

The Biopharmaceutics Classification System divides drugs and drug candidates into four classes based on their solubility and permeability characteristics [1]. Soluble and permeable drugs are designated as Class I compounds, whose oral bioavailability is only limited by the rate at which they reach the appropriate sites of absorption in the gastrointestinal tract. Class II drugs are poorly soluble but permeable through the gut, which suggests that their oral absorption is limited by solubility and dissolution rate of the drugs. Class III compounds are soluble but poorly permeable, so their oral bioavailability is limited by the barrier properties of the gastrointestinal tract. Finally, Class IV compounds are both insoluble and poorly permeable and therefore combine the limitations of both Class II and III compounds. Importantly, an extremely high proportion of currently marketed drugs and drug candidates are not sufficiently water-soluble to achieve therapeutic concentrations [2]. The toxicity and lack of efficacy of drugs and drug candidates are also major causes of failure during drug development [3].

Cyclodextrins (CDs) interact with appropriately sized drug molecules to form inclusion complexes. Generally, these noncovalent inclusion complexes offer a variety of advantages over the uncomplexed drugs. Therefore, CDs are rapidly becoming accepted in pharmaceutics as functional excipients that increase the aqueous solubility, bioavailability, and stability of poorly water-soluble drugs. CDs can also reduce or prevent gastrointestinal and ocular irritation, reduce or eliminate unpleasant smells or tastes, prevent drug–drug or drug–additive interactions, and convert easily crystallized drugs into microcrystalline or amorphous powders. The low toxicity and low immunogenicity of CDs make their use in pharmaceutical formulation development extremely attractive. To extend the application of CDs, several CD derivatives have been developed, including hydroxypropyl-CD (HP-CD), sulfobutyl ether-CD sodium salt (SBE-CD), and methylated-CD (M-CD). Many CD polymers have also been developed. However, the preparation of these polymers usually involves toxic substances, which strongly restrict their clinical applications. Therefore, the application of CD polymers in pharmaceutics is not discussed here. According to the Biopharmaceutics Classification System, CD modification is most applicable to Class II and IV compounds, and it is hoped that CDs can alter the properties of these compounds so that they behave like Class I drugs.

The use of CDs and their derivatives is currently widespread in formulations that involve the oral, parenteral, nasal, pulmonary, and skin delivery of drugs. In 1976, the world's first CD-containing pharmaceutical product, prostaglandin E2–β-CD (Prostarmon E^{TM} sublingual tablets), was marketed in Japan [4]. In 1988, the first European CD-based pharmaceutical product, piroxicam–β-CD (Brexin® tablets), was marketed by Chiesi Farmaceutici (Italy) [5,6]. In 1997, the first U.S.-approved product, itraconazole–2-HP-β-CD oral solution (Sporanox®), was introduced [4]. Today, about 40 different drugs are currently marketed as solid or solution-based CD-complexed formulations (Table 5.1) [5].

Table 5.1. Some marketed pharmaceutical products containing CDs [5].

Drug/CD	Trade name	Formulation	Country
α-Cyclodextrin			
Alprostadil (PGE$_1$)	Prostavastin, Rigidur	i.v. solution	Japan, Europe, USA
OP-1206	Opalmon	Tablet	Japan
Cefotiam-hexetil HCl	Pansporin T	Tablet	Japan
β-Cyclodextrin			
Benexate HCl	Ulgut, Lonmiel	Capsule	Japan
Cephalosporin (ME 1207)	Meiact	Tablet	Japan
Chlordiazepoxide	Transillium	Tablet	Argentina
Dexamethasone	Glymesason	Ointment	Japan
Diphenhydramin HCl, Chlortheophyllin	Stada-Travel	Chewing tablet	Europe
Iodine	Mena-Gargle	Solution	Japan
Nicotine	Nicorette, Nicogum	Sublingual tablet, chewing gum	Europe
Nimesulide	Nimedex	Tablet	Europe
Nitroglycerin	Nitropen	Sublingual tablet	Japan
Omeprazole	Omebeta	Tablet	Europe
PGE$_2$	Prostarmon E	Sublingual tablet	Japan
Piroxicam	Brexin, Flogene, Cicladon	Tablet, suppository, Liquid	Europe, Brazil
Tiaprofenic acid	Surgamyl	Tablet	Europe
2-Hydroxypropyl-β-cyclodextrin			
Cisapride	Propulsid	Suppository	Europe
Itraconazole	Sporanox	Oral and i.v. solutions	Europe, USA
Mitomycin	Mitozytrex	i.v. infusion	Europe, USA
Methylated β-cyclodextrin			
Cloramphenicol	Clorocil	Eye drop solution	Europe
17β-Estradiol	Aerodiol	Nasal spray	Europe

(Continued)

Table 5.1.   (*Continued*)

Drug/CD	Trade name	Formulation	Country
Sulfobutylether β-cyclodextrin			
Voriconazole	Vfend	i.v. solution	Europe, USA
Ziprasidone mesylate	Geodon, Zeldox	IM solution	Europe, USA
2-Hydroxypropyl-γ-cyclodextrin			
Diclofenac sodium	Voltaren	Eye drop solution	Europe
Tc-99 Teoboroxime	Cardiotec	i.v. solution	USA

5.2 Roles of CDs in traditional drug delivery systems

5.2.1 *Control of solid-state properties of drugs*

Many solid compounds exist in different crystalline states, such as amorphous or crystalline states. Some compounds have different crystalline forms, a phenomenon known as "polymorphism". Crystalline modifications significantly affect various pharmaceutical properties, including the solubility, dissolution rate, stability, and bioavailability of drugs. Consequently, it is essential to control the crystallization and polymorphic transitions of solid drugs. It has been proved that crystalline drugs, such as nifedipine [7, 8], chloramphenicol palmitate (CPP) [9], tolbutamide [10, 11], and chlorpropamide [12], can be converted to an amorphous form by complexation with amorphous HP-β-CD. To prevent the crystal growth of nifedipine during storage, HP-β-CD was employed and compared with polyvinylpyrrolidone (PVP). Amorphous nifedipine powders were prepared by spray-drying with HP-β-CD or PVP, and their crystal-growing behavior at accelerated storage conditions of 60°C and 75% relative humidity (RH) was examined. Although PVP initially retarded the crystallization of nifedipine, it failed to control the increase of crystal size after prolonged storage at the accelerated storage conditions, resulting in a remarkable decrease in dissolution rate in water. In contrast, a relatively fine and uniform size of nifedipine crystals was maintained in the HP-β-CD system at accelerated storage conditions. The enhanced dissolution observed for all the HP-β-CD systems in a dissolution medium were clearly reflected in the *in vivo* absorption of nifedipine following oral administration to dogs. CPP was also converted to an amorphous complex when spray-dried with HP-β-CD, and no crystallization of CPP was observed for at least 2 months under storage conditions of 50°C and 50% RH. In contrast, a metastable form of CPP, Form subB, was predominantly formed by spray-drying in the absence of HP-β-CD, and Form subB was easily converted to stable Form B, with a half-life of only 0.5 h under the storage conditions described above. The dissolution rate of the CPP–HP-β-CD complex in a dissolution medium was much faster than those of the CPP polymorphs, which was confirmed in the *in vivo* absorption behavior of CPP after its oral administration to dogs. Furthermore, the effects of storage on the crystallization, dissolution, and absorption of tolbutamide in its amorphous HP-β-CD complex were investigated, and compared with those of a PVP solid dispersion. During accelerated storage conditions of 60°C and 75% RH, tolbutamide in the amorphous PVP dispersion formed a stable from of crystal, whereas tolbutamide in the HP-β-CD complex formed a metastable form of crystal. The dissolution rate of tolbutamide from both the HP-β-CD complex and PVP dispersion was significantly faster than that of the drug alone. However, the dissolution rate of tolbutamide from the PVP dispersion decreased markedly during storage, because slow-dissolving stable crystals formed. In contrast, the dissolution rate from the HP-β-CD complex decreased slightly because fast-dissolving metastable crystals formed. These *in vitro* dissolution characteristics were clearly reflected in the *in vivo* oral absorption behavior of tolbutamide and its glucose plasma levels after its oral administration to dogs. Furthermore, the crystallization of drugs from CD complexes during storage seems to differ from that of drugs associated with PVP. Similar results were obtained for the chlorpropamide–HP-β-CD complex system. Chlorpropamide has a metastable polymorph, Form C, and a stable polymorph, Form A. In the solid state, chlorpropamide was converted to the amorphous complex by spray-drying it with HP-β-CD in a molar ratio of 1:1. During storage, Form C was rapidly converted to Form A. However, Form C crystals were slowly produced from the amorphous HP-β-CD complex, and the further conversion of the resultant Form C to Form A was markedly suppressed in the HP-β-CD matrix. The dissolution rates were in the order of HP-β-CD complex > Form C > Form A, indicating that HP-β-CD increased the dissolution rate of chlorpropamide. HP-β-CD also increased the blood level of chlorpropamide after its oral administration to dogs. To sum up, these studies suggest that HP-β-CD is useful not only for converting crystalline drugs to amorphous substances but also in maintaining the fast dissolution rate of the drugs over a long period.

5.2.2 *Enhancing solubility*

The majority of pharmaceutically active agents are insufficiently soluble in water, and traditional formulation systems for insoluble drugs involve a combination of organic solvents, surfactants, and extreme pH conditions, which often cause irritation or other adverse reactions. It is widely reported that CDs can interact with poorly water-soluble compounds to increase their apparent solubility [13]. This increased apparent solubility allows the development of solution-based dosage forms, such as parenteral formulations and oral liquids. Increasing the apparent solubility of a drug also increases the drug dissolution rate, as defined by the Noyes–Whitney equation, and can increase the oral bioavailability of compounds whose oral bioavailability is limited by their weak solubility or slow dissolution rate [14, 15]. The solubilizing effect is dependent on the type of CD. For example, the solubilizing effects of SBE-β-CD on 22 different poorly water-soluble drugs were compared with those of β-CD and 2,6-di-O-methyl-β-CD (DM-β-CD) [13]. SBE-β-CD was generally a more effective solubilizer for poorly water-soluble drugs than β-CD, but SBE-β-CD was not as effective as DM-β-CD (Table 5.2).

The mechanism underlying this solubilization is based on the ability of CDs to form noncovalent dynamic inclusion complexes in solution. The central CD cavity provides a lipophilic microenvironment, and suitably sized drug molecules can enter the cavity, thus forming drug–CD complexes. Therefore, no covalent bonds are formed or broken during the formation of the drug–CD complex, and

Table 5.2. Effect of CDs on the solubility of slightly soluble or insoluble drugs in water at 25°C [13].

Drug	Solubility in water (μg/mL)	Solubility in 15 mg/ml CD Solution (μg/ml)		
		SBE-β-CD	β-CD	DM-β-CD
Testosterone propionate	1.2	1946 (1622)	6.9 (5.7)	3460 (2845)
Nifedipine	5.4	14.8 (2.7)	10.8 (2.0)	27.7 (5.1)
Benzthiazid	5.5	41.2 (7.5)	44.7 (8.1)	73.4 (13)
Indomethacin	6.9	29.5 (4.3)	12.9 (1.9)	56.0 (8.0)
Digitoxin	9.7	1470 (152)	533 (55)	4469 (460)
Progesterone	11.7	1161 (99)	12.5 (2.1)	2173 (186)
Piroxicam	13.6	73.4 (5.4)	61.5 (4.5)	80.1 (5.9)
Polythiaride	14.6	74 8 (5.1)	71.1 (4.9)	187.4 (1 3)
Acetohexamid	17.4	82.7 (4.8)	55.9 (3.2)	130.7 (7.5)
Griseofulvin	21.2	83.0 (3.9)	26.5 (1.7)	33.5 (1.3)
Spironolactone	24.2	2018 (83)	2174 (90)	4323 (180)
Sulfadirnethoxin	28.2	87.8 (3.1)	121 (4.3)	210.6 (7.5)
FI iirbiprofen	31.3	914.3 (29)	81 4 (2.6)	1721 (55)
Furosernide	32.5	108.5 (3.3)	67.3 (2.1)	121.2 (3.7)
Dipoxin	33.8	1925 (57)	5226 (154)	5193 (154)
17α-Mcthyltestosterone	34.0	1314 (39)	204.7 (6.0)	2340 (69)
Sulfadiazine	90.0	180.2 (2.0)	360.7 (4.0)	347.6 (3.9)
Tolbutarnid	105.3	264.1 (2.5)	206.5 (2.0)	470.8 (4.5)
Ketoprofen	133.0	866.8 (6.5)	1151 (8.7)	1756 (13)
Trichlormcthiazide	148.9	156.8 (1.1)	245.7 (1.7)	255.1 (1.7)
Carhamazepine	167.0	821.5 (4.9)	1108 (6.6)	1975 (12)
Sulfathiazol	441.7	1352 (3.1)	2910 (6.6)	2892 (6.5)

the complexes are readily dissociated in aqueous solution without the formation or breakage of covalent bonds [4]. The forces that drive the formation of the inclusion complexes include the release of enthalpy and water molecules from the cavity, electrostatic interactions, van der Waals interactions, hydrophobic interactions, hydrogen bonding, the release of conformational strain, and charge-transfer interactions [16, 17]. All these forces are relatively weak, allowing the free drug molecules in solution to be in rapid equilibrium with the drug molecules bound within the CD cavity. Therefore, the rates for the formation and dissociation of drug–CD complexes are very close to the diffusion-controlled limits, and drug–CD complexes are continuously being formed and broken apart [18]. CDs and drugs can also form noninclusion complexes because the hydroxyl groups on the outer surface of the CD molecule may form hydrogen bonds with the drug. For instance, α-CD forms both inclusion and non-inclusion complexes with α,ω-dicarboxylic acids, and the two types of complexes coexist in aqueous solution [19]. It has also been reported that when a 1:1 acridine–DM-β-CD inclusion complex forms a noninclusion complex with a second acridine molecule, a complex with a molar ratio of acridine to DM-β-CD of 2:1 is formed [20]. This could explain why the equilibrium constant of complex formation is sometimes concentration-dependent and why its numerical value is frequently dependent on the method used. However, the inclusion type of guest–host CD complexes is probably much more common than noninclusion CD complexes [4].

Moreover, both CDs and CD complexes can self-associate to form water-soluble aggregates, which can further solubilize the drug (Fig. 5.1) [21–26]. In general, negligible amounts of aggregates are formed in pure CD solutions, but aggregation is greatly enhanced by the formation of inclusion complexes, and the extent of aggregation increases as the CD concentration increases [26]. The diameter of the aggregates formed is frequently less than $\sim$300 nm, but visible aggregates can be formed under certain conditions [26]. These aggregates are confirmed to be good solubilizers for ibuprofen and diflunisal [21].

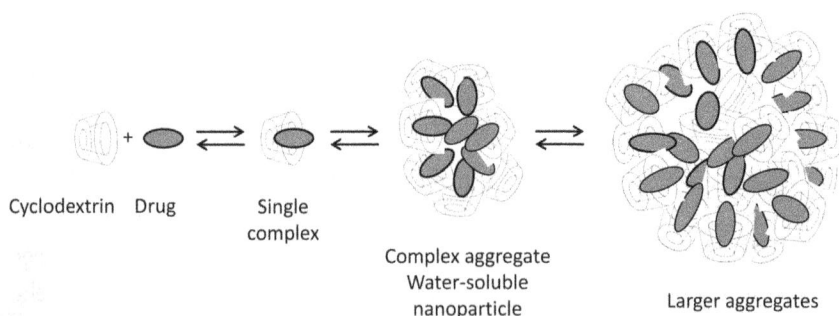

Cyclodextrin   Drug   Single
                           complex

                                   Complex aggregate
                                   Water-soluble
                                   nanoparticle   Larger aggregates

Fig. 5.1.   Formation of complex aggregates in aqueous solution [26].

Another solubilizing attribute of CDs is their ability to form and stabilize supersaturated drug solutions. It has been reported that supersaturated solutions of pancratistatin can be prepared in HP-β-CD solutions [27]. HP-β-CD was also found to be effective in enhancing the stability of supersaturated solutions of ibuprofen [28]. However, the supersaturation should be formed in specific conditions. For example, stable supersatured solutions of silatecan 7-*t*-butyldimethylsilyl-10-hydroxycamptothecin (DB-67) were obtained by mixing a concentrated alkaline aqueous solution of DB-67 car-boxylate with an acidified 22.2% (w/v) SBE-β-CD solution [29]. However, slow addition of DB-67/DMSO into 22.2% (w/v) SBE-β-CD failed to yield stable supersaturated solutions due to precipitation. Supersaturation of the drugs forms, because nucleation and crystal growth of drugs can be inhibited by CDs. The mechanisms by which CDs inhibit crystal nucleation and growth are complicated. CDs solubilize drugs by forming inclusion complexes, and uncomplexed CD may also contribute to the solubilization of drugs. This increases the chemical potential of the drug in solution, increases the apparent saturation solubility, and reduces the extent of supersaturation [4, 28]. However, this effect is not often able to account for the effect on the supersaturated solution, based on the magnitude of the changes in apparent solubility. Another contributing factor may be the interaction between CD and the growing crystal, via either (1) its hydrogen bonding to sites associated with crystal growth; (2) its accumulation in the unstirred water layer, resulting in increased viscosity and hence diffusional resistance; or (3) the complexation of CD with the drug monomers, inhibiting efficient mass transfer at the interface [4].

5.2.3 *Increasing stability*

Some drugs and drug candidates are unstable and readily damaged by oxygen, heat, light, or enzymes. The formation of inclusion complexes with CDs can protect these drugs. Although the CD cavity is open at both ends, steric hindrance protects the guest drug molecules from oxygen, heat, light, and enzymes. Furthermore, because van der Waals forces, hydrophobic forces, hydrogen bonding, etc. act between the guest drug molecules and the host CD molecules, the drugs must overcome a larger energy barrier to volatilize. Therefore, complexed drugs are more stable than free drugs. For example, the doxycycline–HP-β-CD inclusion complex is much more stable than free doxycycline both in solution and as a lyophilized solid, because the unstable site at 6-CH$_3$ on the doxycycline molecule is protected in the hydrophobic cavity of HP-β-CD [30].

5.2.4 *Adjusting the release rate*

Inclusions formed by different kinds of CDs can be used to adjust the release rate of a drug, allowing immediate release, prolonged release, or delayed release. Immediate-release formulations of analgesics, antipyretics, coronary vasodilators, etc. are particularly useful in emergency situations [31]. When poorly water-soluble drugs are complexed with CDs with excellent water solubility, such as HP-β-CD and SBE-β-CD, the dissolution rate of the drugs will be substantially increased. In this case, the complex is suitable for developing drugs to treat acute diseases. Therefore, this immediate-release complex is often used in injections, sublingual tablets, or orally disintegrating tablets. For instance, the chewing-gum tablet containing a fenoprofen calcium–β-CD inclusion complex displays 90% drug dissolution in 5 min, whereas the chewing-gum tablet containing fenoprofen calcium displays 90% drug dissolution in 30 min [32].

Most prolonged-release preparations are intended to maintain a constant release rate so that the final drug concentration in the blood is maintained within a relatively stable range as long as possible. These formulations have many advantages, including a reduced frequency of dosing, prolonged drug efficacy, improved patient compliance, and a reduced risk of fluctuations in the plasma concentration of the drug. Hydrophobic CDs are able to prolong the release of water-soluble drugs. For example, peracylated β-CDs with different chains were used to design a sustained-release formulation for molsidomine, because this drug is water-soluble and has a short biological half-life [33]. The release rate of molsidomine was markedly retarded by complexation with peracylated β-CDs, particularly perbutanoyl–β-CD. When the complexes were orally administered to beagle dogs, perbutanoyl–β-CD suppressed the peak plasma level of molsidomine and maintained a sufficient drug level for a long period, whereas the other derivatives with shorter or longer chains proved inadequate [33]. A similar result was reported for metformin hydrochloride, a highly water-soluble antihyperglycaemic agent with a short biological half-life [34]. When the inclusion complex containing triacetyl-β-CD and metformin hydrochloride was dispersed in a Eudragit®L100-55–chitosan polymeric matrix, about 30% of the drug was released after 2 h at the gastric pH, and the amount of drug released within the following 3 h exceeded 90% in jejunal fluid [34].

Because some drugs are preferentially released in the intestinal tract, these drugs are expected to show delayed release after their oral administration. The stomach is acidic, with a low pH, whereas the intestine has a much higher pH than the stomach. Therefore, under the control of pH, a delayed-release dosage form, which passes from the stomach into the upper small intestine, would experience increased drug release. CDs can be used in drug formulations to achieve this delayed release. For example, carboxymethyl-ethyl-β-cyclodextrin (CME-β-CD) was developed to display pH-dependent solubility for use in the selective dissolution of drug–CD complexes [35–37]. CME-β-CD has limited solubility under acidic conditions, such as in the stomach, whereas the solubility of the complex increases as the pH passes through the pK (3.7) of CME-β-CD [35]. Studies of diltiazem demonstrated that the release rate of the drug from a compressed tablet containing the inclusion complex was significantly retarded in solutions at low pH and increased with increasing pH, and this was reflected in the blood levels of the drug in dogs after its oral administration [35]. Specifically, diltiazem absorption was slower at high gastric acidity ($T_{max} = 4.0 \pm 0.5$ h) than at low gastric acidity ($T_{max} = 2.3 \pm 0.2$ h) (Fig. 5.2) [36]. Molsidomine absorption from tablets containing the molsidomine–CME-β-CD complex was investigated in gastric-acidity-controlled dogs in the fasted and fed states [37]. Molsidomine absorption was significantly retarded under high gastric acidity compared with that under low gastric acidity. The delayed-absorption effect under high gastric acidity was more pronounced under fasting conditions. Clearly, CME-β-CD has

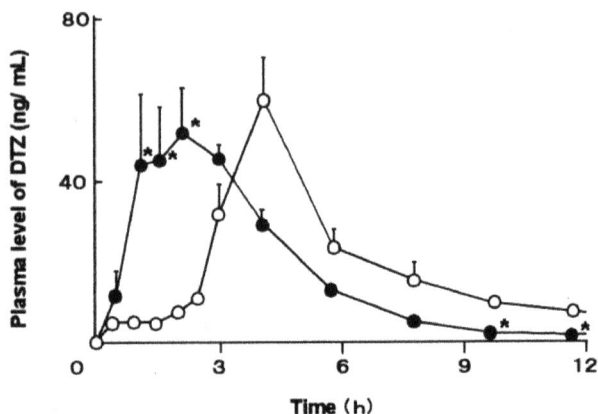

Fig. 5.2.   Plasma levels of diltiazem after the oral administration of tablets containing CME-β-CD complex (equivalent to diltiazem at 30 mg/body) in gastric acidity-controlled dogs; o high gastric acidity dogs; • low gastric acidity dogs. Each value represents the mean ± standard error of four dogs. *, $p < 0.05$ vs. high gastric acidity dogs [36].

potential utility as a delayed-release carrier, where the release rate of the water-soluble drug is reduced in the stomach and increases at its main site of absorption in the small intestine [38].

5.2.5 *Increasing bioavailability*

Cyclodextrins have been extensively used in pharmaceutical formulations to enhance oral bioavailability. The increase in bioavailability is usually expressed as a change in the area under the plasma concentration vs. time curve (AUC), a change in the time to reach the maximum plasma level of a given compound (T_{max}), and/or the maximum plasma level achieved (C_{max}) [39]. The increases in AUC brought about by utilizing CDs with drugs are associated with both increases and reductions in T_{max} and C_{max}. This implies that when the overall exposure to a compound is increased by CD in the dosage form, the accompanying kinetics of drug absorption in the intestine may increase or decrease. A reduction in T_{max} with the inclusion of CD is logical because CD accelerates the dissolution kinetics. An increase in T_{max} could be attributable to the precipitation of a low-solubility administered complex or a reduction of the free drug concentration in the intestinal lumen because of binding to CD. An increase in C_{max} is logical because CDs accelerate drug dissolution, and higher levels of drug in solution in the intestine typically result in greater drug absorption. It has been noted that when an increase in the blood plasma levels of a given compound is not desired, CDs can be used to reduce the required dose. Evidently, there is no clear correlation between the magnitude of the increase in C_{max} or the change in T_{max} and the magnitude of the increase in AUC. In some reports [40, 41], the pharmacodynamics are also used in addition to or instead of the pharmacokinetic changes brought about by the inclusion of CD in a dosage form, which demonstrates that CDs can improve the efficacy of a given dose of a therapeutic agent.

The vast majority of studies in which CDs are used to enhance the bioavailability of a drug use a complex rather than a physical mixture of the given drug and CDs. When a physical mixture is used, it typically results in a significantly lower level or complete lack of improvement in bioavailability. The increase in AUC is approximately twofold higher for complexes than for physical mixtures of single compounds and is often accompanied by a greater reduction in T_{max} and a greater increase in C_{max} [39]. This is probably related to the reduced enhancement in the dissolution kinetics observed when a physical mixture is used rather than a complex. Therefore, if CDs are to constitute a suitable technology for improving bioavailability, a complex must form between the drug and the CD.

The main reason for the increase in bioavailability when a drug is included within CD in the dosage form is the consequent increases in solubility and the dissolution kinetics. As stated above, when the rate-limiting step in drug absorption is dissolution rather than permeation through the intestinal membrane, CDs can increase the drug's bioavailability by increasing its solubility and dissolution kinetics. As shown in Fig. 5.3, once a solid drug is delivered to the gastrointestinal tract as either the free drug, a physical mixture with CD, or a complex with CD, its dissolution and permeation across the intestinal membrane must occur prior to its absorption [39]. If the dissolution kinetics is limited in the overall absorption process,

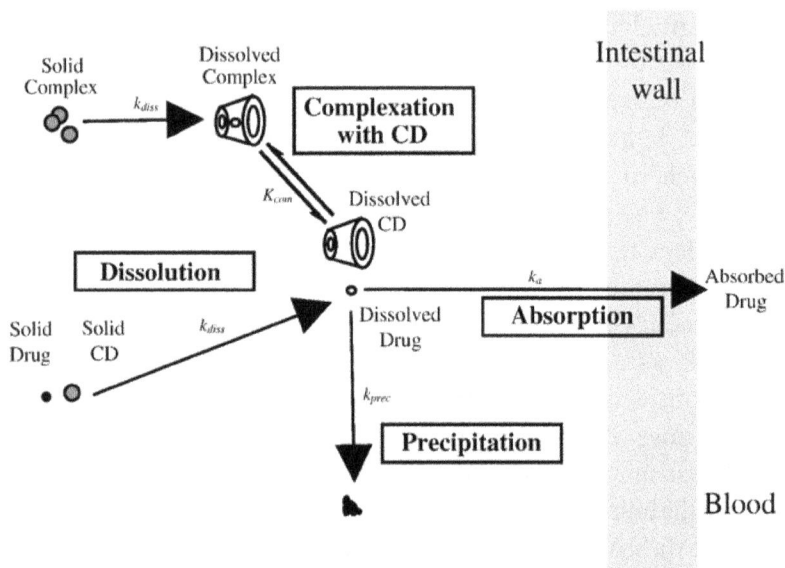

Fig. 5.3.   Diagram of the processes in the intestinal drug-delivery system when a drug is dosed either as a physical mixture or a complex with CD [39].

improving the dissolution kinetics by administering the drug in a physical mixture with CD or as a CD complex will probably increase its bioavailability. The improvement in the dissolution kinetics induced by physically mixing the drug and CD is related to the fact that the dissolution of a drug involves the flux of both the free drug and the complexed drug away from the drug particle surface if CD is present in the immediate environment [42]. However, improvements in the solubility and dissolution kinetics are not always accompanied by an improvement in bioavailability. For example, both β-CD and HP-β-CD significantly improved the solubility and dissolution rate of rutin, but only HP-β-CD increased its bioavailability [43]. Therefore, it is important to identify the rate-limiting step in the absorption of a compound when CD is not present and then assess the ability of CD to influence this rate-limiting step [39].

The bioavailability of some compounds is increased by CDs because their stability is enhanced by their complexation with CDs. The stabilizing effect of CD on these drugs generally enhances their bioavailability. On one hand, the stability of a drug can be enhanced in the dosage form itself by inhibiting polymorphic transitions, as stated above. On the other hand, binding to CD can reduce the reactivity of a drug by shielding the drug molecule from attack and/or by changing the chemical and physical stability of the molecule itself through conformational changes. Thus, once the drug is actually administered, its complexation with CD can prevent its degradation by pH or enzymes, thus improving its stability *in vivo*. Thus, if a compound is susceptible to degradation at the intestinal pH or by intestinal enzymes, complexation with CD can help to maintain the intestinal drug concentration and its absorption through the intestinal membrane. For example, when salbutamol was complexed with β-CD or *tert*-butyl-β-CD (TB-β-CD), its bioavailability increased 1.7-fold or 4.6-fold, respectively, relative to that of the drug administered alone [44]. This was attributed to its reduced biotransformation (mainly glucuronidation) in the intestine. *In vitro* tests demonstrated that complexation reduced the dissolution rate of this highly soluble drug, especially when it was complexed with the hydrophobic TB-β-CD. Such protection may be particularly useful in the oral delivery of peptides or proteins, which are considerably susceptible to enzymatic degradation in the intestine. However, because many studies have involved low-solubility compounds, it is difficult to determine whether the increase in bioavailability is attributable to the effect of the CD on drug dissolution, degradation, or both.

The bioavailability of drugs may also be increased by the interaction between the CDs and biological membranes and the possible associated changes in permeability [39]. Free CDs may form inclusion complexes with membrane components or remove membrane components, thus improving the permeability of the intestine. The nature and extent of the interaction between CDs and the intestinal membrane are not completely understood but appear to depend on the type of CD [39]. For instance, α-CD selectively released phospholipids from rat intestinal membranes, whereas β-CD mainly released cholesterol [45]. The removal of the cholesterol from the membrane by β-CD led to an increased permeability of sulfanilic acid in the transcellular route rather than in the paracellular route. α-CD has also been shown to remove some fatty acids from nasal mucosa and to enhance the nasal absorption of leuprolide acetate in rats and dogs [46]. P-glycoprotein (P-gp) and multidrug-resistance-associated protein 2, which significantly decrease the oral bioavailability of the drugs, are also affected by CDs, and the effect depends on the CD type [46, 47]. For instance, among various CDs, DM-β-CD significantly inhibited these efflux proteins of Caco-2 cell monolayers [47]. It was hypothesized that the inhibition of P-gp by DM-β-CD is attributable to the solubilizing effect of DM-β-CD on cholesterol, an abundant component of the caveolae where P-gp localizes. In general, possible interactions between the solubilizing capacities of CDs, their inhibition of degradation, and/or the changes they induce in intestinal properties make it difficult to develop quantitative guidelines for the use of CDs to enhance bioavailability [39].

5.2.6 *Reducing toxic effects and irritation*

A CD inclusion complex can increase the solubility of a drug and promote its absorption, thereby reducing the dose of the drug required and reducing its toxic effects and irritation. For example, complexation with CDs improved the *in vitro* antiviral activity of ganciclovir against human cytomegalovirus and therefore allowed the administration of lower doses and reduced the toxic effects of ganciclovir [48]. Some irritant or bitter drugs also enter the cavity of CDs upon complexation. Therefore, when the inclusion formulation is administered, direct contact between the drug molecules and the mucosa or the surface of the bowel can be avoided. In this way, the irritation caused by drugs is reduced and bad odors are masked. Another theoretical possibility is that CDs interact with the gate-keeper proteins of the taste buds and paralyze them. When a drug must be dissolved in the oral cavity and rapidly released, masking the taste of the drug using a CD-inclusion complex is particularly important.

5.3 Use of CDs with new classes of drug delivery systems

Several new classes of drug delivery systems encapsulate drugs in colloidal structures to facilitate their desirable pharmacokinetics and drug interactions to target specific tissues. Such systems include hydrogels, liposomes, particles, emulsions, and polymeric micelles. However, these systems usually have some shortcomings and deficiencies, including the suboptimal regulation of drug release and the limited loading of hydrophobic drugs. On the other hand, if a drug–CD complex is directly injected, the complex may rapidly dissociate because the included drug is either diluted or displaced by other blood components. The released drug is then distributed and metabolized as the free drug. Therefore, the degree to which a drug's *in vivo* pharmacokinetics can be improved by its complexation with CD might be limited. Furthermore, after the intravenous administration of naked drug–CD complexes, the interaction between the CD molecules and the cholesterol of red blood cells may cause the drug to be displaced from the complex, with subsequent hematological toxicity. Currently, the combination of CDs and these systems appears to overcome their shortcomings and deficiencies or combine their merits.

5.3.1 *CDs in hydrogels*

Structurally, a hydrogel consists of a relatively small amount of solid components, mostly hydrophilic polymers, dispersed in a large volume of water. These hydrophilic polymers swell in aqueous media without dissolution and form three-dimensional structures. Hydrogels can be prepared with a wide range of sizes, from macrogels to nanogels. Hydrogels can absorb large amounts of water or biological fluids, and hydrophilic drugs can dissolve in water or biological fluids. Therefore, hydrogels have been developed as drug delivery systems (Fig. 5.4) [3]. However, hydrogels are not suitable for hydrophobic drugs because of the large amount of water or biological fluids in the system. Moreover, hydrogels are sometimes unable to control the release of drugs well.

The combination of CDs and hydrogels can overcome the limited loading of hydrophobic drugs in hydrogels and improve the control of drug delivery. CDs increase the network/water partitioning of drugs, whereas the network structure protects the CD inclusion complexes from rapid dilution once in contact with physiological fluids. Hydrogels containing CDs can be classified according to the freedom of the CDs to move inside the network: physically dispersed

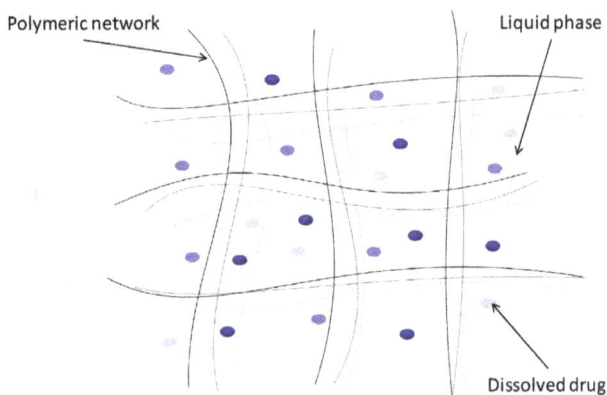

Fig. 5.4.   Hydrogels used as drug delivery systems [3].

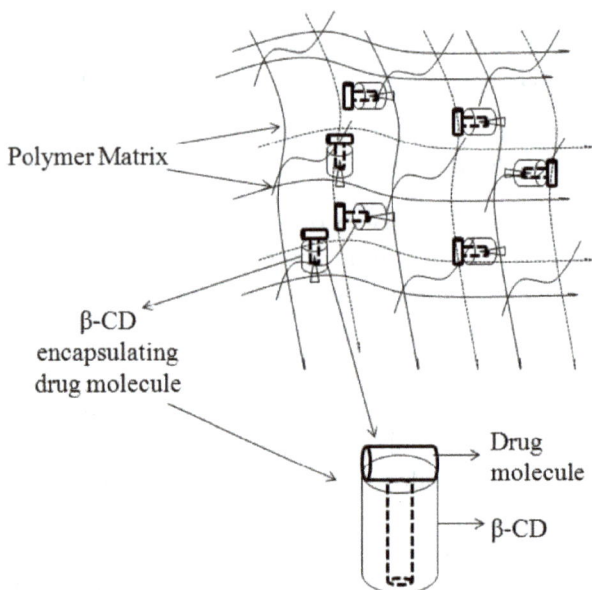

Fig. 5.5.   Schematic representation of hydrogel matrix containing the β-CD–drug inclusion complex [49].

CDs or CDs covalently bonded to the polymeric network. Complex formation allows the formulation of hydrophobic drugs as dissolved solutes in hydrophilic polymer systems (Fig. 5.5) [49]. Furthermore, physically dispersed CDs potentially enhance the release of the drug from polymeric systems by increasing the concentration of diffusible species within the matrix if the drug is loaded at doses above saturation [50]. On the contrary, they can reduce the free drug concentration and slow the release rate if the drug concentration is below its solubility or if the drug forms highly stable complexes. An example of this behavior is the hydroxypropyl methylcellulose gel containing HP-β-CD or M-β-CD used for the nasal release of melatonin [51]. When a low concentration (1%) of CD was added, the nasal penetration of the drug was enhanced. Conversely, at higher CD concentrations (5–10%), nasal penetration decreased. Therefore, different factors must be taken into consideration, including the drug–CD–polymer interactions, medium properties, medium composition, and steric hindrance, because they can influence the release of drugs from hydrogels containing CDs. Free CDs can also modify the drug-release properties of a hydrogel, even if the drug and CDs do not form inclusion complexes. The incorporation of soluble CDs can promote the formation of channels in the hydrogel matrix, accelerating drug release. In contrast, CDs may reduce drug release for one to three main reasons [38]. Firstly, upon complexation with CDs, the molecular weight of drugs is effectively increased, and their diffusivity is reduced. Secondly, CDs reduce the concentration of diffusible species by forming poorly soluble complexes. Thirdly, CDs act as cross-linking agents and reduce the polymer mesh size.

5.3.2 *CDs in liposomes*

Liposomes are microscopic vesicles in which an aqueous solution is entirely enclosed by one or more membrane bilayers. Liposomes are usually made of natural, biodegradable, nontoxic, and nonimmuno-genic lipid molecules, such as phosphatidylcholine and cholesterol [52]. They can encapsulate hydrophilic substances in the aqueous core and lipophilic substances in the membrane. Therefore, they have been used as carriers to deliver active molecules. Liposomal formulations can enhance the pharmacokinetics and pharmacodynamics of the encapsulated drugs, because these formulations induce rapid uptake and long retention by the target tissues [53]. Liposomes can also prevent local irritation and reduce drug toxicity [54, 55]. However,

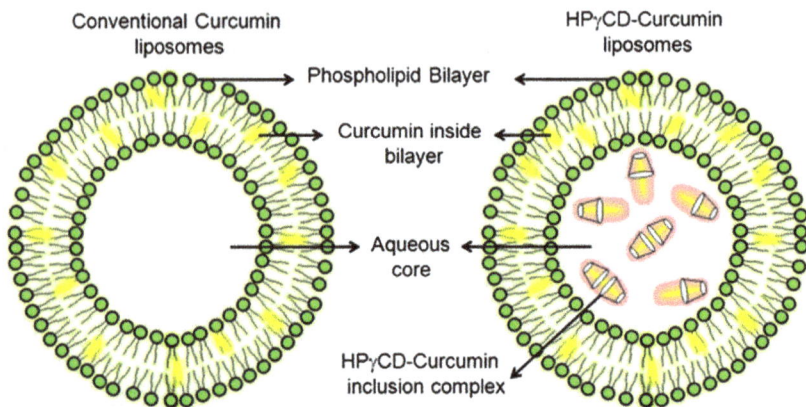

Fig. 5.6.   Schematic representation of conventional and HP-γ-CD–curcumin liposomes [59].

the efficiency of entrapment of a hydrophobic drug by the lipid bilayer usually relies on the mass ratio of the drug to the lipid [56]. It is difficult to achieve high entrapment efficiency in lipid bilayers because the space offered by lipid bilayers is limited, and large amounts of hydrophobic drug molecules can destabilize liposomal bilayers. Moreover, drugs incorporated in the lipid bilayers of liposomes are likely to be more rapidly released than those in the aqueous cores [57].

To ensure the stable encapsulation of lipophilic drugs, an approach was proposed in 1994, wherein drug–CD inclusion complexes are inserted into liposomes (Fig. 5.6) [58–60]. This approach combines the relative advantages of both carriers in a single "drug-in-CD-in-liposome" (DCL) system. The entrapment of the water-soluble CD–drug inclusion complexes in liposomes should allow the accommodation of insoluble drugs in the aqueous phase of the vesicles. The DCL system would also allow the more prolonged release of soluble drugs [60]. Therefore, coupling both delivery systems by encapsulating the CD–drug inclusion complex into liposomes was proposed to circumvent the drawbacks of each separate system. A number of bioactive drugs have been encapsulated in DCLs, including curcumin [59], ketoprofen [61], and β-lapachone [62].

The DCL approach has been shown to improve the stability of liposomal vesicles in most cases. However, CDs, particularly the methylated derivatives, are known to remove the lipid components (especially cholesterol) from cell membranes and liposomes by forming inclusion complexes with them. Therefore, it is possible that lipids can enter the CD cavity to replace the drug during or after the preparation of the DCLs. This could destabilize the bilayers to some extent, allowing the partial or complete leakage of the drug content from the liposomes and altering the vesicle size. Rapid drug release would eliminate the benefits of DCLs. The interaction between CDs and lipids depends on the type of CD, the lipid composition, and the type of liposome. Therefore, it is essential to improve the stability of DCLs by selecting suitable CDs, lipids, and types of liposomes. Another strategy to improve the stability of DCLs is coating them with polymers.

DCLs have been characterized in terms of their morphology, size, encapsulation efficiency, and rate of drug release. These characteristics can be influenced by various factors, including the liposome composition, the method used to prepare the liposomes and drug–CD inclusion complexes, and the type and concentration of CD used. Many studies have demonstrated that the incorporation of a drug into liposomes in the form of a drug–CD inclusion complex improves the encapsulation efficiency. The hydrophobic drug is directly and rapidly released from the lipid bilayer because the hydrophobic drug is predominantly entrapped in the lipid bilayer. Encapsulating drug–CD complexes within the inner aqueous phase of liposomes could modulate the release of the drug molecules. In most cases, DCLs present a more prolonged release profile than liposomes. For example, the release profiles of three different usnic-acid-loaded formulations (inclusion complex, liposome, and DCL) were compared [63]. Almost 50% of the usnic acid in the inclusion complex was released at 3 h, corresponding to an initial burst effect, followed by a linear release of 80% at 11 h, reaching 99.1% in 30 h. The release of usnic acid from the liposomes reached 30% within 7 h, 55% in 24 h, and 70% in 72 h, whereas its release from DCLs was significantly slower (32.5% at 24 h) than that from liposomes (50% at 20.5 h and 65% at 33 h).

Two routes may explain the release of drugs from DCLs [56]. In one route, the drug–CD inclusion complexes are transported from the inner aqueous phase to the lipid bilayer and are then released as intact complexes. In the other route, the free drug is released from the inclusion complexes in the inner aqueous phase of the DCLs and is transported from the inner aqueous phase to the lipid bilayer. The release rate is affected by the constant of the inclusion complex. Regardless of the route the release process follows, the lipid bilayer barrier must be overcome first. Therefore, DCLs may present more prolonged release profiles than liposomes because there are more barriers to the diffusion of the drug from DCLs.

5.3.3 *CDs in microparticles or nanoparticles*

As is implied by the name, microparticles are small particles with the diameter ranging from 1 to 1000 μm [64]. Microparticles have several advantages in drug delivery, including a predictable and reproducible gastrointestinal transit time, being little influenced by the presence of food, together with minor local irritation and/or adverse effects, and consistent drug absorption [65]. Nanoparticles are colloidal polymeric carriers with sizes in the 1–1000 nm range [64]. Nanoparticle drug delivery systems have specific advantages based on their size and stability: (1) they have a large surface area, which allows better contact with biological membranes; (2) they can pass through the smallest capillary vessels and avoid rapid clearance by phagocytes; and (3) they can penetrate cells and tissue gaps to arrive at the target organ [66]. Therefore, they can also be suitably functionalized to provide site-specific delivery, local drug release, and/or enhanced intracellular delivery [64]. However, their very low drug-loading capacity and limited entrapment efficiency restrict the uses of nanoparticles.

Analogous to hydrogel and liposome systems, the combination of CD complexation and loading the complex into a microparticle or nanoparticle system has been proposed [64, 67]. The aim of such combined strategies is to simultaneously exploit both the solubilizing effect of CD complexation and the carrier properties of microparticle or nanoparticles, thus achieving more effective drug delivery devices [64]. It has been demonstrated that amphiphilic CDs spontaneously form nanoparticles when a nanoprecipitation technique is used [67].

CDs can be particularly useful for hydrophobic drugs. For example, the ketoprofen–CD complex was successfully entrapped in polymeric microparticles intended for colonic delivery, and permeation studies confirmed the clearly increased permeation of the drug formulated as a CD complex in microparticles, demonstrating the synergistic effects of CD and chitosan in improving drug permeation [68]. A pharmacokinetic study also demonstrated that after the oral administration of daidzein-loaded CD inclusion complexes in poly(lactide-co-glycolide) acid (PLGA) nanoparticles to rats at a dose of 10 mg/kg, the relative bioavailability of the drug was enhanced 8.85-fold compared with that of a daidzein suspension [69]. Nanoparticles charged with a flurbiprofen–HP-β-CD complex showed significantly higher *in vivo* topical anti-inflammatory efficacy than nanoparticles loaded with flurbiprofen alone [70]. The drug release rate from nanoparticles is strongly influenced by the kind of CD used. For example, the percentage of the drug released at 24 h varied from 16% (plain oxaprozin-loaded PLGA nanoparticles) to 50% (oxaprozin–β-CD-loaded PLGA nanoparticles) and to 100% (oxaprozin–M-β-CD-loaded PLGA nanoparticles) [71].

5.3.4 *CDs in microemulsions or nanoemulsions*

Microemulsions are thermodynamically stable, transparent, isotropic, low-viscosity colloidal dispersions consisting of microdomains of oil and/or water stabilized by an interfacial film of alternating surfactant and cosurfactant molecules [72]. They include swollen micellar (oil-in-water, O/W), reverse micellar (water-in-oil, W/O), and bicontinuous structures. Microemulsions offer interesting drug delivery vehicles, given their ease of preparation, long-term stability, and high solubilization capacity for both hydrophilic and lipophilic drugs. Therefore, the utility of microemulsions, particularly O/W microemulsions, in dermal, oral, ocular, and parenteral delivery is continuously being explored. One major drawback of microemulsions is that their formation requires a high surfactant concentration, usually about 20 wt% or higher, which may cause toxicity when used in pharmaceutical applications [73]. Furthermore, the low proportion of the oil phase in O/W microemulsions limits the incorporation of lipophilic drugs. On the other hand, O/W nanoemulsions contain small oil droplets (diameter < 500 nm) dispersed within a watery continuous phase, with each oil droplet surrounded by a protective coating of emulsifier molecules. The small droplet size makes them stable against sedimentation and creaming for a long time, thus increasing the overall stability of the emulsion. The capacity of nanoemulsions to dissolve large quantities of hydrophobic drugs, together with their mutual compatibility and ability to protect the drugs from hydrolysis and enzymatic degradation, make them ideal vehicles for parenteral transport [74]. However, nanoemulsions are thermodynamically unstable, and the droplet size tends to increase during storage.

Therefore, the combined use of microemulsions or nanoemulsions and CDs has been proposed to overcome their drawbacks [73, 75]. The addition of β-CD or M-β-CD improved the proportion of the oil phase in soya-oil-based O/W microemulsions, thus enhancing the incorporation of the lipophilic drugs, sulfamerazine and indomethacin [76]. Increased permeation of both drugs (by 90% for sulfamerazine and by 75% for indomethacin) was also observed when microemulsions with β-CD were used, indicating that combining microemulsions with CDs (mainly M-β-CD) may improve the diffusion of lipophilic drugs through the unstirred water layers of biological membranes. Cryotransmission electron microscopy (Cryo TEM) images showed that the addition of γ-CD to lecithin-based or sucrose-stearate-based nanoemulsions markedly improved the homogeneity of the formulation (Fig. 5.7) [77]. Furthermore, the addition of CD markedly increased the skin permeation of progesterone and fludrocortisone acetate relative to the corresponding systems without CD. These results suggest that CD was incorporated into the interfacial films of the nanoemulsions, which might alter the drug release rate and increase the stability of the droplet microstructure.

5.4 Toxicological considerations of CDs

Because CDs have a large number of hydrogen donor and acceptor molecules, high molecular weights, and very low octanol/water

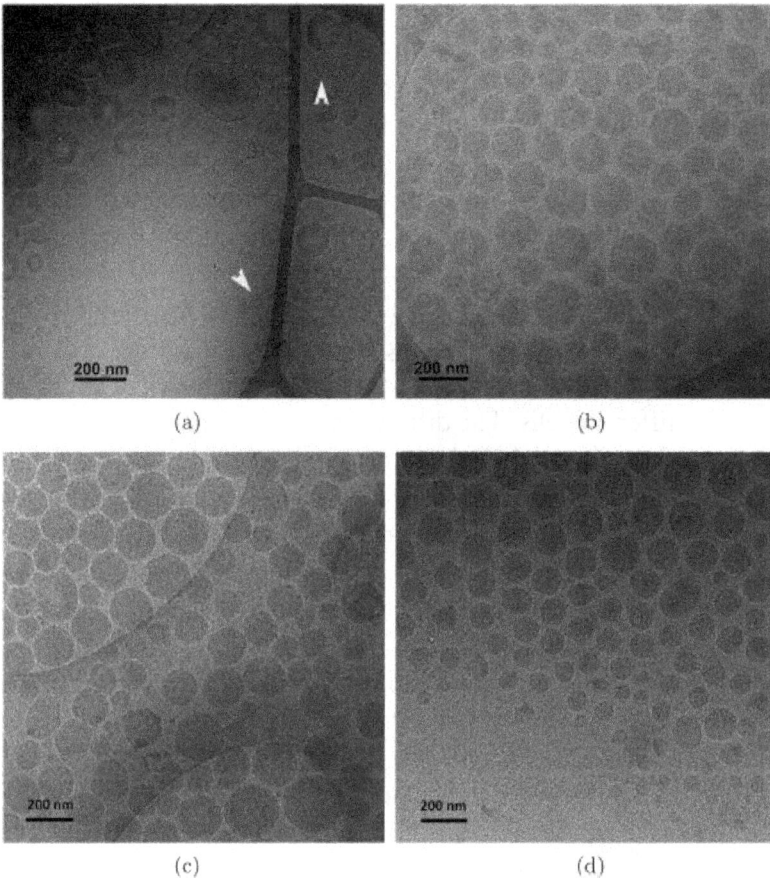

Fig. 5.7. Cryo TEM photographs of the microstructures of blank nanoemulsions (a) with 2.5% lecithin E-80 and (b) with 2.5% sucrose stearate S-970 as sole emulsifying agent. The corresponding formulations containing additional γ-CD are shown below for lecithin (c) as well as for sucrose stearate (d) [77].

partition coefficients, they usually do not readily permeate biological membranes [78, 79]. Therefore, most CDs cannot be absorbed from the gastrointestinal tract in their intact form, and negligible amounts of hydrophilic CDs and drug–CD complexes can permeate lipophilic membranes, such as the gastrointestinal mucosa and skin [80]. Studies have demonstrated that CDs are practically nontoxic when administered orally because they are not absorbed from the gastrointestinal tract [80]. After intravenous (i.v.) injection, CDs are almost exclusively eliminated through the kidneys. Therefore, patients who suffer severe renal insufficiency should not be administered CDs via i.v. injection. The toxicity of CDs may be mainly revealed in their toxicity on cells, and the toxicity profiles of different CDs depend on their type. For instance, the hemolytic effect of CDs on human erythrocytes in phosphate-buffered saline is in the order of methylated β-CDs > β-CD > HP-β-CD > α-CD > γ-CD > HP-γ-CD > SBE-β-CD [80]. This may be attributable to the different abilities of the CDs to bind to or extract cholesterol from cell membranes [80]. Moreover, the same CD usually displays different toxicity in different cells. The difference in the toxicities of the CDs in different cell types may be related to the different lipid compositions of the plasma membranes. CDs can also form inclusion complexes with other compounds within membranes, such as biliary salts and proteins, and these compounds differ in different cell membranes.

Besides the CD type, the toxicity profiles of CDs also depend on the route of administration. Natural α-CD and β-CD have been used in marketed oral dosage formulations (Table 5.1), which confirm their safety after oral administration. Oral administration of α-CD is well tolerated and is not associated with significant adverse effects [81, 82]. α-CD has been also used in a marketed i.v. solution, suggesting its safety after i.v. administration. However, when administered by the parental route, especially by the i.v. route, β-CD can destabilize red blood cell membranes by forming inclusion complexes with some of their constituents, leading to hemolysis [83]. Therefore, β-CD cannot be given parenterally because of its low aqueous solubility and adverse effects [4]. However, it is essentially nontoxic when given orally. When orally administered, doses of 0.7–0.8 g/kg/day β-CD in rats and ~2 g/kg/day in dogs were shown to be nontoxic [84]. The metabolism of γ-CD closely resembles that of starch and linear dextrin, because human salivary and pancreatic α-amylases are able to hydrolyze γ-CD [85]. The oral administration of 8 g γ-CD or 8 g maltodextrin to humans showed no difference in gastrointestinal tolerance [86]. Furthermore, the three natural CDs are accepted as food additives, which confirm their safety to humans through oral administration.

The acceptable daily oral intakes for humans are 1.4 g for α-CD, 0.35 g for β-CD, 10 g for γ-CD [87].

HP-β-CD, a hydrophilic CD derivative, is much more water-soluble and more toxicologically benign than natural β-CD [88], which is confirmed by the use of HP-β-CD in marketed i.v. drug formulations (Table 5.1). Clinical studies have shown that HP-β-CD is tolerated well and is safe in the majority of patients receiving HP-β-CD at a daily oral dose of 4–8 g for at least 2 weeks [80]. However, higher oral daily doses of 16–24 g resulted in an increased incidence of soft stools and diarrhea when given to volunteers for 14 days [88]. Based on these results, HP-β-CD is considered to be nontoxic (at least for 14 days), if the daily dose is <16 g. In an intravenous dosing study, single doses up to 3 g were found to have no measurable effect on kidney function and were well tolerated by all volunteers [88]. Moreover, following a 1-week intravenous study at a single dose level of 1 g, no adverse effects were reported. SBE-β-CD is also included in several marketed products (Table 5.1), which confirms its safety at low doses. However, few published references on the toxicological potential of SBE-β-CD are available. The available toxicological information on HP-γ-CD is also limited, but HP-γ-CD is also used in two products, an eye drop solution formulation and a parenteral diagnostic product (Table 5.1).

Methylated CDs, the lipophilic CD derivatives, are also used in marketed products, including an eye drop solution and a nasal spray. Scanning electron microscopic observation also confirmed that a 10-min exposure to 1.3% DM-β-CD caused no significant changes in the surface morphology of rat nasal mucosa [89]. However, the i.v. administration of DM-β-CD to rats at the low dose (300 mg/kg, 50 mg/kg × 6) led to an increase of glutamate-pyruvate transaminase and glutamate-oxaloacetate transaminase, indicating some hepatic disorder [80]. This increase in enzymes may impart to DM-β-CD a relatively low LD_{50} value, which is 220 mg/kg in rats. The systemic toxicity of DM-β-CD may be due to its higher surface activity as well as greater ability to interact with endogenous lipids, which characteristically limits its parenteral uses. Only limited data are available on the oral safety of the methylated CDs; for example, the oral administration of DM-β-CD in aqueous solution at doses up to 3 g/kg to mice resulted in no toxic symptoms [80]. However, the oral administration of M-β-CD is currently limited by its potential toxicity [4].

5.5 Regulatory status of CDs

As the intense study of CDs progresses, the regulatory status of CDs continues to evolve. α-CD and β-CD have their own monographs in the current European Pharmacopoeia (EP), British Pharmacopoeia (BP), United States Pharmacopoeia (USP), and Japanese Pharmaceutical Codex (JPC) [4, 15, 90]. γ-CD is also referenced in the USP and JPC. They are classified as excipients, rather than as parts of drug substances, although opinions about this classification are divided. Among the CD derivatives, only HP-β-CD has a legal status and its own pharmacopoeial monographs in the current EP and BP [90]. The problem with derivatives mainly arises from the heterogeneity of the final product. Furthermore, α-CD, β-CD, and γ-CD have been introduced as food additives into the "generally regarded as safe" (GRAS) list of the U.S. Food and Drug Administration (FDA) in 2004, 2001, and 2000, respectively [4]. Besides the three natural CDs, HP-β-CD, HP-γ-CD, M-β-CD, and SBE-β-CD are also cited in the FDA's list of pharmaceutical ingredients.

References

1. Loftsson T. (2002). Cyclodextrins and the biopharmaceutics classification system of drugs. *Journal of Inclusion Phenomena and Macrocyclic Chemistry*, 44, pp. 63–67.
2. Lipinski CA. (2002). Poor aqueous solubility — an industry wide problem in drug discovery. *American Pharmaceutical Review*, 5, pp. 82–85.
3. Moya-Ortega MD, Alvarez-Lorenzo C, Concheiro A, Loftsson T. (2012). Cyclodextrin-based nanogels for pharmaceutical and biomedical applications. *International Journal of Pharmaceutics*, 428, pp. 152–163.
4. Brewster ME, Loftsson T. (2007). Cyclodextrins as pharmaceutical solubilizers. *Advanced Drug Delivery Reviews*, 59, pp. 645–666.
5. Loftsson T, Duchene D. (2007). Cyclodextrins and their pharmaceutical applications. *International Journal of Pharmaceutics*, 329, pp. 1–11.
6. Loftsson T, Brewster ME. (2012). Cyclodextrins as functional excipients: Methods to enhance complexation efficiency. *Journal of Pharmaceutical Sciences*, 101, pp. 3019–3032.

7. Uekama K, Ikegami K, Weng Z, Horiuchi Y, Hirayama F. (1992). Inhibitory effect of 2-hydroxypropyl-β-cyclodextrin on crystal-growth of nifedipine during storage: Superior dissolution and oral bioavailability compared with polyvinylpyrrolidone k-30. *Journal of Pharmacy and Pharmacology*, 44, pp. 73–78.

8. Hirayama F, Wang Z, Uekama K. (1994). Effect of 2-hydroxypropyl-β-cyclodextrin on crystallization and polymorphic transition of nifedipine in solid state. *Pharmaceutical Research*, 11, pp. 1766–1770.

9. Hirayama F, Usami M, Kimura K, Uekama K. (1997). Crystallization and polymorphic transition behavior of chloramphenicol palmitate in 2-hydroxypropyl-β-cyclodextrin matrix. *European Journal of Pharmaceutical Sciences*, 5, pp. 23–30.

10. Kimura K, Hirayama F, Arima H, Uekama K. (1999). Solid-state ^{13}C nuclear magnetic resonance spectroscopic study on amorphous solid complexes of tolbutamide with 2-hydroxypropyl-α-and -β-cyclodextrins. *Pharmaceutical Research*, 16, pp. 1729–1734.

11. Kimura K, Hirayama F, Arima H, Uekama K. (2000). Effects of aging on crystallization, dissolution and absorption characteristics of amorphous tolbutamide-2-hydroxypropyl-β-cycylodextrin complex. *Chemical and Pharmaceutical Bulletin* (Tokyo), 48, pp. 646–650.

12. Sonoda Y, Hirayama F, Arima H, Uekama K. (2004). Effects of 2-hydroxypropyl-β-cyclodextrin on polymorphic transition of chlorpropamide in various conditions: Temperature, humidity and moulding pressure. *Journal of Inclusion Phenomena and Macrocyclic Chemistry*, 50, pp. 73–77.

13. Ueda H, Ou D, Endo T, Tomono K, Nagai T. (1998). Evaluation of a sulfobutyl ether β-cyclodextrin as a solubilizing agent for several drugs. *Drug Development and Industrial Pharmacy*, 24, pp. 863–867.

14. Vromans H, Eissens A, Lerk C. (1989). Mechanism of dissolution of drug-cyclodextrin complexes: a pragmatic approach. *Acta Pharmaceutica Technologica*, 35, pp. 250–255.

15. Challa R, Ahuja A, Ali J, Khar RK. (2005). Cyclodextrins in drug delivery: An updated review. *AAPS PharmSciTech*, 6, pp. E329–E357.

16. Loftsson T, Berw ME. (1996). Pharmaceutical applications of cyclodextrins. 1. Drug solubilization and stabilization. *Journal of Pharmaceutical Sciences*, 85, pp. 1017–1025.

17. Gelb RI, Schwartz LM, Cardelino B, Fuhrman HS, Johnson RF, Laufer DF. (1981). Binding mechanisms in cyclohexaamylose complexes. *Journal of the American Chemical Society*, 103, pp. 1750–1757.

18. Stella VJ, Rao VM, Zannou EA, Zia V. (1999). Mechanisms of drug release from cyclodextrin complexes. *Advanced Drug Delivery Reviews*, 36, pp. 3–16.

19. Gabelica V, Galic N, Pauw ED. (2002). On the specificity of cyclodextrin complexes detected by electrospray mass spectrometry. *Journal of the American Society for Mass Spectrometry*, 13, pp. 946–953.

20. Correia I, Bezzenine N, Ronzani N, Platzer N, Beloeil JC, Doan BT, (2002). Study of inclusion complexes of acridine with β- and (2,6-di-o-methyl)-β-cyclodextrin

by use of solubility diagrams and nmr spectroscopy. *Journal of Physical Organic Chemistry*, 15, pp. 647–659.

21. Magnusdottir A, Masson M, Loftsson T. (2002). Self association and cyclodextrin solubilization of nsaids. *Journal of Inclusion Phenomena and Macrocyclic Chemistry*, 44, pp. 213–218.

22. Loftsson T, Magnúsdóttir A, Másson M, Sigurjónsdóttir JF. (2002). Self-association and cyclodextrin solubilization of drugs. *Journal of Pharmaceutical Sciences*, 91, pp. 2307–2316.

23. Loftsson T, Másson M, Brewster ME. (2004). Self-association of cyclodextrins and cyclodextrin complexes. *Journal of Pharmaceutical Sciences*, 93, pp. 1091–1099.

24. Auzély-Velty R. (2011). Self-assembling polysaccharide systems based on cyclodextrin complexation: Synthesis, properties and potential applications in the biomaterials field. *Comptes Rendus Chimie*, 14, pp. 167–177.

25. Messner M, Kurkov SV, Brewster ME, Jansook P, Loftsson T. (2011). Self-assembly of cyclodextrin complexes: Aggregation of hydrocortisone/cyclodextrin complexes. *International Journal of Pharmaceutics*, 407, pp. 174–183.

26. Ryzhakov A, Do Thi T, Stappaerts J, Bertoletti L, Kimpe K, Sa Couto AR, Saokham P, Van den Mooter G, Augustijns P, Somsen GW, Kurkov S, Inghelbrecht S, Arien A, Jimidar MI, Schrijnemakers K, Loftsson T. (2016). Self-assembly of cyclodextrins and their complexes in aqueous solutions. *Journal of Pharmaceutical Sciences*, 105, pp. 2256–2269.

27. Torres-Labandeira JJ, Davignon P, Pitha J. (1991). Oversaturated solutions of drug in hydroxypropylcyclodextrins: Parenteral preparation of pancratistatin. *Journal of Pharmaceutical Sciences*, 80, pp. 384–386.

28. Iervolino M, Raghavan S, Hadgraft J. (2000). Membrane penetration enhancement of ibuprofen using supersaturation. *International Journal of Pharmaceutics*, 198, pp. 229–238.

29. Xiang TX, Anderson BD. (2002). Stable supersaturated aqueous solutions of silatecan 7-t-butyldimethylsilyl-10-hydroxycamptothecin via chemical conversion in the presence of a chemically modified β-cyclodextrin. *Pharmaceutical Research*, 19, pp. 1215–1222.

30. Zhang H, Chen M, He Z, Wang Z, Zhang M, He Z, Wan Q, Liang D, Repka MA, Wu C. (2013). Molecular modeling-based inclusion mechanism and stability studies of doxycycline and hydroxypropyl-β-cyclodextrin complex for ophthalmic delivery. *AAPS PharmSciTech*, 14, pp. 10–18.

31. Hirayama F, Uekama K. (1999). Cyclodextrin-based controlled drug release system. *Advanced Drug Delivery Reviews*, 36, pp. 125–141.

32. El Assassy AE, Amin MM, Abdelbarya AA. (2012). Immediate release three-layered chewing gum tablets of fenoprofen calcium: Preparation, optimization and bioavailability studies in healthy human volunteers. *Drug Development and Industrial Pharmacy*, 38, pp. 603–615.

33. Uekama K, Horikawa T, Yamanaka M, Hirayama F. (1994). Peracylated β-cyclodextrins as novel sustained-release carriers for a water-soluble drug, molsidomine. *Journal of Pharmacy and Pharmacology*, 46, pp. 714–717.

34. Corti G, Cirri M, Maestrelli F, Mennini N, Mura P. (2008). Sustained-release matrix tablets of metformin hydrochloride in combination with triacetyl-beta-cyclodextrin. *European Journal of Pharmaceutics and Biopharmaceutics*, 68, pp. 303–309.
35. Uekama K, Horiuchi Y, Irie T, Hirayama F. (1989). O-carboxymethyl-o-ethylcyclomalthoheptaose as a delayed-release-type drug carrier: Improvement of the oral bioavailability of diltiazem in the dog. *Carbohydrate Research*, 192, pp. 323–330.
36. Uekama K, Horikawa T, Horiuchi Y, Hirayama F. (1993). *In vitro* and *in vivo* evaluation of delayed-release behavior of diltiazem from its o-carboxymethyl-o-ethyl-β-cyclodextrin complex. *Journal of Controlled Release*, 25, pp. 99–106.
37. Horikawa T, Hirayama F, Uekama K. (1995).*In vivo* and *in vitro* correlation for delayed-release behaviour of a molsidomine/o-carboxymethyl-o-ethyl-β-cyclodextrin complex in gastric acidity-controlled dogs. *The Journal of Pharmacy and Pharmacology*, 47, pp. 124–127.
38. Bibby DC, Davies NM, Tucker IG. (2000). Mechanisms by which cyclodextrins modify drug release from polymeric drug delivery systems. *International Journal of Pharmaceutics*, 197, pp. 1–11.
39. Carrier RL, Miller LA, Ahmed I. (2007). The utility of cyclodextrins for enhancing oral bioavailability. *Journal of Controlled Release*, 123, pp. 78–99.
40. Vila-Jato JL, Blanco J, Torres JJ. (1987). Biopharmaceutical aspects of the glibornuride-β-cyclodextrin inclusion compound. *STP Pharma*, 3, pp. 28–32.
41. Choi HG, Lee BJ, Han JH, Lee MK, Park KM, Yong CS, Rhee JD, Kim YB, Kim CK. (2001). Terfenadine-β-cyclodextrin inclusion complex with antihistaminic activity enhancement. *Drug Development and Industrial Pharmacy*, 27, pp. 857–862.
42. Bekers O, Uijtendaal EV, Beijnen JH, Buit A, Underberg WJM. (1991). Cyclodextrins in the pharmaceutical field. *Drug Development and Industrial Pharmacy*, 15, pp. 1503–1549.
43. Miyake K, Arima H, Hirayama F, Yamamoto M, Horikawa T, Sumiyoshi H, Noda S, Uekama K. (2000). Improvement of solubility and oral bioavailability of rutin by complexation with 2-hydroxypropyl-β-cyclodextrin. *Pharmaceutical Development and Technology*, 5 pp. 399–407.
44. Hirayama F, Horikawa T, Yamanaka M, Uekama K. (1995). Enhanced bioavailability and reduced metabolism of salbutamol by perbutanoyl β-cyclodextrin after oral administration in dogs. *Pharmacy and Pharmacology Communications*, 1, pp. 517–520.
45. Nakanishi K, Nadai T, Masada M, Miyajima K. (1992). Effect of cyclodextrins on biological membrane. II. Mechanism of enhancement on the intestinal absorption of non-absorbable drug by cyclodextrins. *Chemical and Pharmaceutical Bulletin* (Tokyo), 40, pp. 1252–1256.
46. Uekama K. (2004). Design and evaluation of cyclodextrin-based drug formulation. *Chemical and Pharmaceutical Bulletin* (Tokyo), 52, pp. 900–915.
47. Yunomae K, Arima H, Hirayama F, Uekama K. (2003). Involvement of cholesterol in the inhibitory effect of dimethyl-β-cyclodextrin on p-glycoprotein and mrp2 function in caco-2 cells. *FEBS Letters*, 536, pp. 225–231.

48. Nicolazzi C, Abdou S, Collomb J, Marsura A, Finance C. (2001). Effect of the complexation with cyclodextrins on the *in vitro* antiviral activity of ganciclovir against human cytomegalovirus. *Bioorganic & Medicinal Chemistry*, 9, pp. 275–282.

49. Singh K, Nair AB, Kumar A, Kumria R. (2011). Novel approaches in formulation and drug delivery using contact lenses. *Journal of Basic and Clinical Pharmacy*, 2, pp. 87–101.

50. Otero-Espinar FJ, Torres-Labandeira JJ, Alvarez-Lorenzo C, Blanco-Méndez J, (2010). Cyclodextrins in drug delivery systems. *Journal of Drug Delivery Science and Technology*, 20, pp. 289–301.

51. Babu RJ, Dayal P, Singh M. (2008). Effect of cyclodextrins on the complexation and nasal permeation of melatonin. *Drug Delivery*, 15, pp. 381–388.

52. Anwekar H, Patel S, Singhai AK. (2011). Liposome as drug carriers. *International Journal of Pharmaceutical and Life Sciences*, 2, pp. 945–951.

53. Qian S, Li C, Zuo Z. (2012). Pharmacokinetics and disposition of various drug loaded liposomes. *Current Drug Metabolism*, 13, pp. 372–395.

54. Budai M, Szogyi M. (2001). Liposomes as drug carrier systems. Preparation, classification, and therapeutic advantages of liposomes. *Acta pharmaceutica Hungarica*, 71, pp. 114–118.

55. Barenholz Y. (2003). Relevancy of drug loading to liposomal formulation therapeutic efficacy. *Journal of Liposome Research*, 13, pp. 1–8.

56. Chen J, Lu WL, Gu W, Lu SS, Chen ZP, Cai BC, Yang XX. (2014). Drug-in-cyclodextrin-in-liposomes: A promising delivery system for hydrophobic drugs. *Expert Opinion on Drug Delivery*, 11, pp. 565–577.

57. Maestrelli F, Gonzalez-Rodriguez ML, Rabasco AM, Ghelardini C, Mura P. (2010). New drug-in cyclodextrin-in deformable liposomes formulations to improve the therapeutic efficacy of local anaesthetics. *International Journal of Pharmaceutics*, 395, pp. 222–231.

58. McCormack B, Gregoriadis G. (1994). Drugs-in-cyclodextrins-in-liposomes: A novel concept in drug delivery. *International Journal of Pharmaceutics*, 112, pp. 249–258.

59. Dhule SS, Penfornis P, Frazier T, Walker R, Feldman J, Tan G, He J, Alb A, John V, Pochampally R. (2012). Curcumin-loaded gamma-cyclodextrin liposomal nanoparticles as delivery vehicles for osteosarcoma. *Nanomedicine: Nanotechnology, Biology, and Medicine*, 8, pp. 440–451.

60. Gharib R, Greige-Gerges H, Fourmentin S, Charcosset C, Auezova L. (2015). Liposomes incorporating cyclodextrin–drug inclusion complexes: Current state of knowledge. *Carbohydrate Polymers*, 129, pp. 175–186.

61. Maestrelli F, Gonzalez-Rodriguez ML, Rabasco AM, Mura P. (2006). Effect of preparation technique on the properties of liposomes encapsulating ketoprofen-cyclodextrin complexes aimed for transdermal delivery. *International Journal of Pharmaceutics*, 312, pp. 53–60.

62. Cavalcanti IM, Mendonca EA, Lira MC, Honrato SB, Camara CA, Amorim RV, Mendes Filho J, Rabello MM, Hernandes MZ, Ayala AP,

Santos-Magalhaes NS. (2011). The encapsulation of beta-lapachone in 2-hydroxypropyl-beta-cyclodextrin inclusion complex into liposomes: A physicochemical evaluation and molecular modeling approach. *European Journal of Pharmaceutical Sciences*, 44, pp. 332–340.

63. Lira MCB, Ferraz MS, Silva DGVC, Cortes ME, Teixeira KI, Caetano NP, Sinisterra RD, Ponchel G, Santos-Magalhães NS. (2009). Inclusion complex of usnic acid with β-cyclodextrin: Characterization and nanoencapsulation into liposomes. *Journal of Inclusion Phenomena and Macrocyclic Chemistry*, 64, pp. 215–224.
64. Maestrelli F, Bragagni M, Mura P. (2016). Advanced formulations for improving therapies with anti-inflammatory or anaesthetic drugs: A review. *Journal of Drug Delivery Science and Technology*, 32, pp. 192–205.
65. Roy P, Shahiwala A. (2009). Multiparticulate formulation approach to pulsatile drug delivery: Current perspectives. *Journal of Controlled Release*, 134, pp. 74–80.
66. Liu Z, Jiao Y, Wang Y, Zhou C, Zhang Z. (2008). Polysaccharides based nanoparticles as drug delivery systems. *Advanced Drug Delivery Reviews*, 60, pp. 1650–1662.
67. Duchene D, Ponchel G, Wouessidjewe D. (1999). Cyclodextrins in targeting application to nanoparticles. Advanced Drug Delivery Reviews, 36, pp. 29–40.
68. Maestrelli F, Zerrouk N, Cirri M, Mennini N, Mura P. (2008). Microspheres for colonic delivery of ketoprofen-hydroxypropyl-beta-cyclodextrin complex. *European Journal of Pharmaceutical Sciences*, 34, pp. 1–11.
69. Ma Y, Zhao X, Li J, Shen Q. (2012). The comparison of different daidzein-plga nanoparticles in increasing its oral bioavailability. *International Journal of Nanomedicine*, 7, pp. 559–570.
70. Vega E, Egea MA, Garduno-Ramirez ML, Garcia ML, Sanchez E, Espina M, Calpena AC. (2013). Flurbiprofen plga-peg nanospheres: Role of hydroxy-beta-cyclodextrin on ex vivo human skin permeation and *in vivo* topical anti-inflammatory efficacy. *Colloids and Surfaces B: Biointerfaces*, 110, pp. 339–346.
71. Muraa P, Maestrelli F, Cecchi M, Bragagni M, Almeid A. (2010). Development of a new delivery system consisting in 'drug–in cyclodextrin–in plga nanoparticles'. *Journal of Microencapsulation: Micro and Nano Carriers*, 27, pp. 479–486.
72. Date AA, Nagarsenker MS. (2008). Parenteral microemulsions: An overview. *International Journal of Pharmaceutics*, 355, pp. 19–30.
73. Kaur K, Kaur Bhatia N, Mehta SK. (2012). Formation of cyclodextrin-stabilized nanoemulsions and microemulsions and exploitation of their solubilization behavior. *RSC Advances*, 2, pp. 8467–8477.
74. Lovelyn C, Attama AA. (2011). Current state of nanoemulsions in drug delivery. *Journal of Biomaterials and Nanobiotechnology*, 2, pp. 626–639.
75. Klang V, Matsko N, Zimmermann AM, Vojnikovic E, Valenta C. (2010). Enhancement of stability and skin permeation by sucrose stearate and cyclodextrins in progesterone nanoemulsions. *International Journal of Pharmaceutics*, 393, pp. 152–160.

76. Aloisio C, Oliveira AG, Longhi M. (2016). Cyclodextrin and meglumine-based microemulsions as a poorly water-soluble drug delivery system. *Journal of Pharmaceutical Sciences*, 105, pp. 2703–2711.
77. Klang V, Matsko N, Raupach K, El-Hagin N, Valenta C. (2011). Development of sucrose stearate-based nanoemulsions and optimisation through gamma-cyclodextrin. *European Journal of Pharmaceutics and Biopharmaceutics*, 79, pp. 58–67.
78. Lipinski CA, Lombardo F, Dominy BW, Feeney PJ. (2001). Experimental and computatorial approaches to estimate solubility and permeability in drug discovery and development settings. *Advanced Drug Delivery Reviews*, 46, pp. 3–26.
79. Loftsson T, Jarho P, Másson M, Järvinen T. (2005). Cyclodextrins in drug delivery. *Expert Opinion Drug Delivery*, 2, pp. 335–351.
80. Irie T, Uekama K. (1997). Pharmaceutical applications of cyclodextrins. III. Toxicological issues and safety evaluation. *Journal of Pharmaceutical Sciences*, 86, pp. 147–162.
81. Lina BA, Bar A. (2004). Subchronic (13-week) oral toxicity study of alpha-cyclodextrin in dogs. *Regulatory Toxicology and Pharmacology*, 39 Suppl 1, pp. S27–S33.
82. Lina BA, Bar A. (2004). Subchronic oral toxicity studies with alpha-cyclodextrin in rats. *Regulatory Toxicology and Pharmacology*, 39 Suppl 1, pp. S14–S26.
83. Perrin JH, Field FP, Hansen DA, Mufson RA, Torosian G. (1978). beta-Cyclodextrin as an aid to peritoneal dialysis. Renal toxicity of beta-cyclodextrin in the rat. *Research Communications in Chemical Pathology & Pharmacology*, 19, pp. 373–376.
84. Bellringer ME, Smith TG, Read R, Gopinath C, Oliver P. (1995). β-Cyclodextrin: 52-week toxicity studies in the rat and dog. *Food and Chemical Toxicology*, 33, pp. 367–376.
85. Munro IC, Newberne PM, Young VR, Bar A. (2004). Safety assessment of gamma-cyclodextrin. *Regulatory Toxicology and Pharmacology*, 39 Suppl 1, pp. S3–S13.
86. Koutsou GA, Storey DM, Bar A. (1999). Gastrointestinal tolerance of gamma-cyclodextrin in humans. *Food Additives and Contaminants*, 16, pp. 313–317.
87. Antlsperger G, Schmid G. (1996). Toxicological comparison of cyclodextrins. In: Szejtli J, Szente L (Eds.), *Proceedings of the Eighth International Symposium on Cyclodextrins*. Dordrecht: Kluwer Academic Publishers, pp. 149–155.
88. Gould S, Scott RC. (2005). 2-hydroxypropyl-beta-cyclodextrin (HP-beta-CD): A toxicology review. *Food and Chemical Toxicology*, 43, pp. 1451–1459.
89. Matsubara K, Abe K, Irie T, Uekama K. (1995). Improvement of nasal bioavailability of luteinizing hormone-releasing hormone agonist, buserelin, by cyclodextrin derivatives in rats. *Journal of Pharmaceutical Sciences*, 84, pp. 1295–1300.
90. Cal K, Centkowska K. (2008). Use of cyclodextrins in topical formulations: Practical aspects. *European Journal of Pharmaceutics and Biopharmaceutics*, 68, pp. 467–478.

6. Applications in Cosmetics

Tao Feng*, Haining Zhuang[†], and Na Yang[‡]

*School of Perfume and Aroma Technology,
Shanghai Institute of Technology,
Shanghai 200235, China

[†]Institute of Edible Fungi,
Shanghai Academy of Agricultural Sciences,
Shanghai 201403, China

[‡]School of Food Science and Technology,
Jiangnan University, Wuxi 214122, China

6.1 Introduction

Nanoencapsulation is an emerging technology that includes the formulation, characterization, manufacture, and application of the encapsulant, equipment, and systems to control the sizes and shapes of particles at the nanometer scale. Here, "nanometer" refers to the range of 1–100 nm, where 1 nm is 10^{-9} m. Cosmetic technologists and scientists recognize that nanotechnology will be the future trend in cosmetic products and will become the most popular and innovative technology. Cosmetic companies use nanoscale particles of functional factors to provide better ultraviolet (UV) protection; deeper skin moisturization and penetration; whitening, antiaging, and slow-release effects; and increased color, smoothing, quality, etc. In 2012, the total global market for cosmetics based on nanoencapsulation was worth US$155.8 million. The properties of these nanoparticles are superior in terms of color, transparency, solubility, and chemical reactivity than those of larger particles, so nanomaterials are attractive to the cosmetics and personal-care industries, and the use of nanoscale materials in cosmetics is widespread.

Different surveys have shown that all the giant cosmetic companies use nanoencapsulation technologies in their various products. For example, the famous cosmetic brand Estee Lauder began to use nanocarriers in 2006, with a series of products including nanoencapsulants. Another famous cosmetics company, L'Oréal, devoted about US$600 million dollars of its US$17 billion dollar revenue to nanopatents, and dozens of "nanosome particles" have been patented. Therefore, L'Oréal ranks sixth among the nanotech patent holders in the USA. Based on the Espacenet database, the top 10 cosmetic companies of the world, ranked from first to tenth in terms of their nanorelated patents, are: L'Oreal, Procter & Gamble, Henkel, Unilever, Kao Corp., Avon, Shiseido, Beiersdorf, Estee Lauder, and Johnson & Johnson.

In the last decade, the cosmetics industry has shown growing interest in cyclic oligosaccharides in its constant search for new materials to improve the characteristics of their cosmetic products. In this context, several advanced technologies have been explored, and among these, a prominent place is occupied by cyclodextrins (CDs). CDs provide many benefits and, in many cases, also release the complexed molecules in a gradual manner over time. For these reasons, they are considered suitable not only for skin treatment products but also for makeup, which stays on the face for many hours, and for perfumes, ensuring a longer life of the fragrance.

In general, CDs can be used in cosmetic formulations to

- increase the water solubility of lipophilic materials;
- reduce or eliminate undesirable odors;
- convert liquid or oily materials to a powder form, improving their handling;
- provide the controlled release of the active ingredients in complex fragrances;
- increase the physical and chemical stability of guest molecules by protecting them from decomposition, oxidation, hydrolysis, or loss by evaporation;

Table 6.1. Cosmetics containing CDs already on the market.

Product type	Product name
Personal-care products	Ointments, cosmetic powders, deodorants, bath preparations, self-tanning creams, perfumes, detergents, antiwrinkle creams
Hair-care products	Permanent wave solutions, shampoos, lotions, conditioning products
Nail products	Lacquer removers

- reduce or prevent skin irritation;
- increase or reduce the absorption of various compounds into the skin; and
- stabilize emulsions and suspensions.

In contrast to starch, CDs and their derivatives are not a nutrient medium for microorganisms, especially yeasts and fibrous fungi. Consequently, the use of preservatives in formulation can be reduced, a further advantage of CDs. The cosmetics industry has begun transforming our knowledge of CDs into products (Table 6.1) and many different cosmetics formulated with CDs are already on the market. These oligosaccharides are also characterized by a high degree of safety when used topically.

Several reports in the literature have reviewed the use of CDs in cosmetics. Most of them have focused on the cosmetic use of empty CDs and inclusion complexes. However, few papers have reviewed the uses of CDs in cosmetics based on their cosmetic functions. These will be discussed in detail below, according to the individual functions of the products.

6.2 Molecular inclusion of sunscreening agents and its application in cosmetics

A safer sunscreen formula that does not irritate the skin or cause phototoxic or photoallergic reactions, and that which can be triggered at the molecular level to penetrate the skin, is urgently required. The search for active substances and efficient combinations and the design of new carriers or vehicles have resulted in the development of novel cosmetic systems, in contrast to the classic types of sunscreens, which are gels or creams. The skin can be protected by applying a sunscreen that comes as a lotion, spray, or another type of topical product. The most effective sunscreens protect against UV-A (near-UV radiation, wavelengths 320–400 nm), UV-B (UV radiation, wavelengths 280–320 nm), and UV-C (far-UV radiation, wavelengths 200–280 nm) radiations [1].

In recent decades, damage to the stratospheric ozone layer has produced a progressive increase in the amount of UV radiation reaching the Earth's surface. It is well documented that exposure to the sun's UV radiation plays a fundamental role in the photodamaging effects on the skin, which include acute inflammatory responses and long-term adverse reactions. As a consequence, there has been an increase in the recognition of the number of illnesses caused by this harmful radiation, which include cutaneous photoaging, skin cancer, and damage to the skin's immunological system [2]. Therefore, the growing public awareness of the dangers of sunlight exposure has increased the widespread use of topical sunscreen preparations. So-called "UV filters", formed by molecular inclusion, have been used as active ingredients in cosmetic sunscreen products to protect the human skin from the damaging radiation of sunlight [3]. These active compounds are organic or inorganic polymers that absorb and/or reflect UV radiation.

When exposed to sunlight, the sunscreen molecule is photoactivated. Ultimately, the excitation energy is dissipated as fluorescence, thermal energy, or phosphorescence or is transferred to the surrounding molecules. The decomposition of the photoinduced sunscreen agent can also occur, reducing its UV-protective capacity, ultimately leading to the accumulation of potentially harmful photolytic products on the skin. The photostability of sunscreening agents encapsulated by β-CD is reported to be considerably greater than that of the same agent uncomplexed or complexed with phospholipids [4].

Sunscreen agents remain the best and safest approach to minimizing the chance of skin cancer and aging [5]. However, recognition of the drawbacks to the use of chemical UV filters in sunscreen products has escalated quickly, probably because the formulations and their properties are still far from perfect. Sunscreen molecules and the other active ingredients found in cosmetic formulations are supposed to stay on the horny layer of the skin, rather than be absorbed by the skin. This may lead, in turn, to systemic problems. Unfortunately, the optimal behavior does not always correspond to reality, and various adverse reactions have been recorded in the last few decades. For instance, phenomenal allergies to ethylhexyl methoxycinnamate have been observed, which are the most widely used chemical filters in sunscreen products, including photoallergy, reactive oxygen species (ROS) production, and estrogenic and nonestrogenic activities [6]. Similar adverse effects have also been reported for other compounds used as chemical UV filters [6].

CDs are cyclic oligosaccharides with an apolar cavity and a hydrophilic outer surface. They can encapsulate appropriately sized lipophilic molecules within their hydrophobic interiors by forming noncovalent complexes [7]. This phenomenon affects the physico-chemical properties of the encapsulated substance in a way that can increase its stability to air and light and its apparent aqueous solubility [8]. A comprehensive review of CDs and their ability to improve both the solubility and stability of active ingredients has been published [9].

Among the various possible strategies to increase sunscreen stability, CDs offer excellent host molecules for sunscreen reagents. Naturally occurring CDs consist of six, seven, and eight glucopyra-nose units, known as α-, β-, and γ-CDs, respectively. Semisynthetic CD derivatives of commercial interest include hydroxypropyl-β-CDs (HP-β-CDs) and hydroxypropyl-γ-CDs (HP-γ-CDs), randomly methylated β-CDs (RM-β-CDs), and sulfobutyl ether-β-CDs (SBE-β-CDs) [10]. These compounds have a three-dimensional (3D) structure shaped like a truncated cone, with a lipophilic central cavity and a hydrophilic outer surface attributable to the presence of hydroxyl moieties.

The presence of a lipophilic cavity allows the accommodation of molecules with the appropriate structural properties [11]. The lipophilic nature of most sunscreen molecules makes them promising candidate guest molecules for CD complexation. CDs have also been used in a number of other applications, for example, as solubility enhancers, in drug delivery systems administered via different routes, and as ingredients of cosmetic products [12]. In this context, the host–guest interaction increases the stability of the guest compound to light and oxygen, reduces its vapor pressure, and modifies some of its physicochemical properties, with clear benefits for the production of sunscreen and skin-care products [12]. Inclusion in CDs is also an appealing strategy for overcoming the well-cited ethical issues of sunscreens. The production of commercial sunscreens, such as *para*-aminobenzoic acid (PABA), 2-ethylhexyl-4-methoxycinnamate (EMC), 4-methylbenzylidene camphor, ethylhexyl dimethyl-PABA, benzophenone-4 (BENZ-4), and menthyl anthranilate, is often based on this inclusion method [13]. A higher UV-filtering capacity and increased stability of the sunscreen molecule have been observed for all these complexes, attenuating any potential skin toxicity. The CD inclusion method can also be readily scaled to an industrial level. In this respect, the complexation of PABA with β-CDs altered the kinetics of PABA-induced thymine dimerization by minimizing the probability of triplet–triplet energy transfer to thymine [14]. The increased photostability of ferulic acid after its the successful inclusion in α-CDs has been reported [15].

Other studies have focused on the skin permeation, retention, and photoprotection of benzophenone-3 (BENZ-3) when added to different CDs [16]. The skin accumulation, permeation, and flux of BENZ-3 were enhanced by HP-β-CDs depending on concentration. The transdermal penetration of B-3 increased at 5% and 10% CD, whereas at 20% CD, the B-3 flux dropped because B-3 was more likely to complex with CD (in large excess) than penetrate the skin [17]. This can be an advantage. Similarly, phenylbenzimidazole sulfonic acid (PBSA), a compound approved for use as a sunscreen in Europe, the USA, and Australia, was formulated as inclusion complexes of α-, β-, and γ-CD derivatives [18]. The stability of the sunscreen was increased at the optimal pH for topical formulations, and the light-induced production of free radicals was suppressed by the complexation of PBSA, especially with HP-β-CD [18].

Butyl methoxy dibenzoylmethane (BMDBM) has also been complexed with β-CDs, HP-β-CDs, and SBE-β-CDs. BMDBM is a sunscreen that is readily photoinactivated, causing cutaneous and ocular irritation, phototoxicity, subcutaneous toxicity, DNA damage, and lipid peroxidation [19]. Significant BMDBM photostabilization and free-radical-scavenging effects were achieved only when BMDBM was combined with HP-β-CDs, whereas its complexation with SBE-β-CDs consistently reduced the penetration of BMDBM in the epidermal layer of human skin [20]. In contrast, this complex suppressed the photodegradation of BMDBM, inhibited its percutaneous penetration, and reduced the *in vivo* numbers of sunburnt cells and the levels of skin edema [21]. This was attributed to the combination of BMDBM inclusion and the physical effects of the sunscreen.

PBSA is a widely used sunscreening agent that absorbs most efficiently in the 290–320 nm region of solar UV radiation [22]. It is listed among the authorized UV filters in Europe [23] and the USA [24]. PBSA is considered photostable, as demonstrated in recent studies [25]. These studies have shown that a variety of free radicals and active oxygen species that cause photoinduced DNA damage *in vitro* are generated by this sunscreen reagent under solar radiation [22]. The preparation and characterization of CD complexes formed with PBSA have been reported. HP-β-CD or RM-β-CD can be used to investigate the effect of a third component (e.g., water-soluble polymers or l-lysine) on the efficiency of complexation. PBSA must be neutralized completely with a suitable base before it is used as a water-soluble UV-B filter because it has very low solubility in both water and oil. The pH of the finished sun-care preparation must also be adjusted to >7.0 to prevent its crystallization [26]. This is a disadvantage because the topical application of the product can alter the physiological (acidic) pH of the skin, causing irritation [27]. It is also important to ensure that no acidic compound is used in these sunscreen formulations because it could precipitate the water-insoluble free acid in PBSA, with a consequent loss of efficacy [26].

Long-term stability studies have shown that HP-β-CD and RM-β-CD are less effective or their activities are almost completely suppressed by the reduction of the sunscreening functions of emulsion preparations (pH 4.0). Moreover, the complexation of PBSA with HP-β-CD markedly reduced its irradiation-induced decomposition in the emulsion vehicle (3.9% degradation of the complex compared with 9.1% degradation of uncomplexed PBSA), whereas RM-β-CD had no significant effect. The formation of free radicals by PBSA upon exposure to simulated sunlight was completely inhibited by its inclusion in the HP-β-CD cavity, which limited its photosensitizing potential [18].

Early studies showed that the formation of ternary complexes between CD, the guest molecule, and a third component can improve the efficiency of complexation [28]. Therefore, the influence of combining HP-β-CD with water-soluble polymers on the aqueous solubility of PBSA was monitored at pH 4.0 after PBSA was activated by heating [28]. Higher proportions of CDs should also reduce the competitive displacement of the guest molecule from the host cavity by various emulsion excipients [29]. A number of studies have shown that an excess of CD reduced the drug absorption through the skin [30]. Consequently, excess CD in test formulations can lessen the percutaneous penetration of PBSA, thereby enhancing its retention on the skin surface where its sunscreening activity is required [31].

Stevenson and Davies [32] and Inbaraj et al. [22] have demonstrated that when exposed to solar UV radiation, PBSA generates a variety of free radicals and ROS, which damage single- and double-stranded DNA in vitro. The radical species (R·) generated during sunscreen irradiation were monitored by detecting the more stable paramagnetic adducts formed according to the principle: 5,5′-dimethyl-1-pyrroline—N-oxide(DMPO) and OH· radicals [33]. Under irradiation in the absence of PBSA, no electron paramagnetic resonance (EPR) signal was detected. The spectrum of the DMPO–OH· adduct clearly indicated that a photosensitization degrading process from PBSA to O_2 occurred [22]. Such adduct is expected to derive from direct trapping of radicals and the very fast decomposition of DMPO–O_2· via an electron transfer process or the singlet oxygen generated in an energy transfer reaction from the photogenerated PBSA triplet state to O_2 [34].

Continuous exposure to external oxidant attack, such as UV radiation, chemicals, or ozone, makes the skin a major target of oxidative stress. To prevent oxidative stress, cells are protected with two major antioxidant defense systems. The first involves enzymes that catalyze the degradation of ROS or reactive nitrogen species (RNS), such as superoxide dismutase or glutathione reductase. The second involves antioxidants, such as Trolox (Sigma-Aldrich Corporation, 2033 Westport Center Dr, St. Louis, MO 63146), that reduce ROS or RNS via oxidoreduction reactions [35]. Trolox (6-hydroxy-2,5,7,8-tetramethylchroman-2-carboxylic acid) is chemically related to vitamin E. The chromanol structure provides its antioxidant activity, whereas the carboxyl group imparts moderate water solubility [35]. Although the antiradical activity of Trolox has been studied extensively, it has several drawbacks, including its poor water solubility, photoinstability, and so on [36, 37]. A quantitative investigation of the complexation of Trolox with HP-β-CD was performed with the Higuchi and Connors method [38]. The photostability of Trolox increased after its complexation with HP-β-CD and the inclusion complexation was significantly influenced by the protonation–deprotonation equilibrium of Trolox. The undissociated Trolox molecule better fits within the CD cavity, whereas dissociated Trolox, which is negatively charged and hydrophilic, interacts with the hydrophobic sites in CD, preventing complex formation.

Trolox absorbs radiation in the UV region, with a peak at 290 nm. However, as previously reported [36], under UVB irradiation, it displays a peak at 273 nm, which is attributed to a photodegradation product, Trolox C quinone. Pure Trolox and the Trolox–HP-β-CD complex were irradiated separately to study the protective effect of CD against the photooxidation of the active molecule. The molecules were irradiated in the presence or absence of TiO_2 to assess the potential photocatalytic effect of this substance. In all media, Trolox was degraded under UVB light with pseudozero-order kinetics, probably by a photooxidation mechanism. In water, the Trolox degradation rate decreased when it was within the HP-β-CD cavity. The photocatalysis induced by TiO_2 was inhibited by the complexation of Trolox with CD, perhaps because the oxide was excluded from the hydrophobic site of the guest molecule because it has a negative charge. This suggests a possible correlation between the host molecule and the pores of lipid membranes. On the basis of this study, it can be said that CD affected the transmembrane transport of Trolox, increasing the rate of its active diffusion. It is well known that CD increases the availability of a drug on the surfaces of biological barriers, maintains hydrophobic drug molecules intact in solution, and unlike some conventional penetration enhancers, such as alcohols and fatty acids, disrupts the lipid layers of biological barriers [39].

The transepidermal permeation and skin uptake of CD-complexed Trolox, determined *in vitro* in porcine ear skin, showed negligible percutaneous permeation within the first 6 h, whereas some Trolox (μg/mL) was observed in all the receiving redox reaction states after 24 h. This enhancement effect could be attributable to the fact that CD increases the availability of the active molecule at the skin surface, consistent with the previously discussed results for diffusion through membranes. The permeation and accumulation of Trolox from micellar solutions of the pure active molecule or its complex with CD were similar, indicating that the enhancement effect of CD was insignificant and can therefore be neglected in such media. To increase the photostability of Trolox, a widely used antioxidant, it was complexed with HP-β-CD. The reduction in the degradation rate was greater in oil-in-water (O/W) emulsions than in hydrophilic media, in which photo-oxidative reactions are usually favored. The Trolox–HP-β-CD complex could play a primary role in protecting the skin from oxidative stress and has great potential utility [40].

Other modified CDs have attracted attention because they have advantages over native CDs [41]. Methyl-β-cyclodextrin dissolves easily and faster in water than β-CD. It forms stable aqueous solutions and, unlike ionic derivatives, it does not significantly increase the osmotic power of the solution, thus streamlining the design process. Complexing sunscreening agents with CDs are a recognized distribution strategy. CDs are perceived as the hosts of a group of guests, and the system studied may allow scientists to put forward the importance of their experimental results. Therefore, when Trolox was complexed with M-β-CD, its stability against UVB irradiation increased. The complexation constants and docking outcomes suggested that this complexation was associated with the pH of the medium. Photodegradation studies of various topical formulae (gel, O/W emulsion, and water–oil–water [W/O/W] emulsion) containing free Trolox or Trolox combined with methyl-β-CD have been reported. In every case, Trolox was degraded according to pseudozero-order kinetics. In the presence of TiO_2, a popular sunscreen known to be a photocatalyzer, the host compound (CD) amplified the photostability of Trolox [42].

Gamma-oryzanol (GO), a combination of ferulic acid esters that has recently attracted much interest, is a natural antioxidant from rice-bran oil, which is usually added to stabilize food and pharmaceutical raw materials. It is also used as a sunscreen in many cosmetic formulae [43]. GO consists of a blend of at least 10phytosteryl ferulates, including cycloartenyl ferulate, 24methylene cycloartanyl ferulate, and campesteryl ferulate [44]. GO has a wide range of biological functions, specifically displaying radical-scavenging characteristics. It is also used in the management of hyperlipoproteinemia and in various cosmetic formulations, including sunscreens [45]. However, its use is impractical because it is rapidly degraded. A proposed method to increase the stability and efficacy of antioxidants is based on the use of supramolecular structures (nanoparticles, CDs, liposomes, and so on). Lipid peroxidation is an oxidative process that occurs via a free-radical chain-reaction mechanism. This process can alter the organoleptic and technological features of oils and fats, thereby reducing their shelf lives. GO abolishes these chain reactions by eliminating free radicals. Therefore, it is used extensively in the cosmetic industry to impede the oxidative degradation of lipid raw materials.

The reaction of CDs with a suitable cross-linking agent allows a nanoporous material, called "nanosponge", to be generated, which displays fascinating properties [46]. Nanosponge can compress many materials so that they can be transferred via aqueous media or, conversely, be eliminated from polluted water. The two major traits of nanosponge are the microscopic spherical shape of its particles and the polarity of the mesh itself, which may be adjusted by the type of cross-linking used, the type of cross-linker, and environmental reactions. These traits allow the addition of other materials, which can be either strongly held or released in a precise way. Nanosponges can also reduce adverse effects and prevent the loss of the guest compound. The effects of the medium (hydroalcoholic solution, micellar solution, gel, or O/W emulsion) have also been studied, and several studies have shown that the photooxidation rate may depend on the polarity of the medium [36, 47]. Because GO is a mixture of ferulic acid esters that undergo photolysis, the photodegradation of this compound when subjected to UV irradiation can be assessed in environments with low polarity (emulsions or micelles), which restrict the photooxidation process. The observed kinetic constraints imply that the photodegradation rate of GO combined with neutral emulsifiers is slower than that of free GO, indicating that the complexation phenomenon provides GO with a physical barrier against UV-induced oxidation [36, 47]. This capacity of neutral emulsifiers to increase the skin uptake of the guest molecule, as detected in both conditions, may be associated with the fact that CDs predominantly improve the infusion of drugs by increasing their solubility at the lipophilic surface of the skin [48, 49]. *In vitro* studies performed with Franz diffusion cells established that the complexation phenomenon does not inhibit the build-up of GO in porcine ear skin. In summary, the creation of NS inclusion complexes containing GO demonstrates the potential utility of *N*-hydroxysuccinimide as a carrier of topically absorbed substances.

In the cosmetics industry, the creation of inclusion CD complexes containing a variety of sunscreen agents has increased the stability and solubility of the cosmetically active ingredients, protecting them from light- and oxidation-induced degradation [50]. Ferulic acid (FA) has a high UV-absorbance property, which suggests that it can protect the skin from sunlight-induced damage. Based on these characteristics, FA has been approved for use as a sunscreen in Japan [51]. Anselmi *et al.* [52] and Centini *et al.* [15] attempted to enhance this application with CDs, which have been shown pharmaceutically to generate stable inclusion complexes by interacting noncovalently with various hydrophobic small compounds. CDs were also used as structures to reduce the percutaneous penetration of popular sunscreens. Because UV filters have various effects on the skin, it is vital that the sunscreen accumulates in the outmost cutaneous layers after its application, with negligible penetration to the deeper skin tissues [20, 53].

The O/W emulsion of a CD–FA complex showed better photo-stability after irradiation with 10 minimal erythemal doses than the free FA emulsion [54]. The photostability of FA increased with the addition of α-CD, mainly because no *cis*-isomers or other degradation products formed after irradiation. However, the inclusion of other forms of CD (β or γ) did not guarantee the same degree of protection of FA [55]. This report stated that the antioxidant activity of the CD–FA complex at all the concentrations tested was less potent than an equimolecular mixture of FA and α-CD or FA alone [55]. These results confirm that FA is firmly encapsulated within the α-CD cavity in the inclusion complex and that its phenol group is less accessible for interaction with peroxyl radicals. Different results were observed when a sunscreen/radical scavenger was included in the α-CD complex. In all cases, the formation of the FA–CD complex reduced the transdermal permeation of FA compared with the same formulations containing free FA [56].

There are many conflicting reports in the literature of the relationship between the complexity and transdermal kinetics of topically applied compounds. Several researchers have shown that CDs enhance penetration, whereas others have reported an increased flux across the skin [48]. It is noteworthy that CDs may enhance transdermal permeation by increasing the solubility of the active ingredient and facilitating its segregation towards the lipophilic surface of the skin. A theoretical model of the complex formation and transdermal permeation of topically applied compounds containing CDs has been developed [48]. In that study, the researchers assessed whether an excess concentration of CD shifts the equilibrium of a drug suspension toward a complex-forming system. Only the free drug in solution penetrated the skin. This explains why the complexation of FA with CD in our study reduced the permeation flux relative to that of free FA. When FA was included in α-CD (CD–FA), the amount of FA retained in the skin layers from gel-type formulations declined noticeably throughout the entire thickness of the skin. In particular, the amount of FA from α-CD–FA, which accumulated in the upper skin layers (15–100 μm), was 1.3–5.7-fold greater than that acquired with α-CD–FA, and the difference increased significantly with increasing depth. When free FA was used in an acidic (pH 6) emulsion, the nonionized FA was divided between the lipid phase of the stratum corneum and the vehicle. In O/W CD–FA, the complex remained in the aqueous phase and interacted directly with the biological barrier, making free FA available at the absorptive surface [57]. Therefore, in both the cases, there was a restrictive stage in the diffusion of FA through the cutaneous barrier, during which FA was partitioned between the emulsion phases and FA was released from the complex [57]. One paper suggested that the CD–FA complex had less antioxidant potency than FA alone [55, 58]. However, the addition of CD protects the molecule from photodegradation, increasing the photostability of FA and consequently ensuring longer-lasting protection. There was no formation of the less-active FA *cis*-isomer or other degradation products.

6.3 Molecular inclusion of skin-whitening agents and its application in cosmetics

Skin whitening (also called "skin bleaching") lightens the complexion with various cosmetic methods [59]. The color of the skin is genetically determined but can be modified by environmental factors. Peoples of different ethnic and cultural backgrounds often have different preferences in skin color. Bronze is a more desirable complexion in Western culture. However, in Eastern populations, a light complexion is the aspiration of every woman because white skin is considered to signify youth and beauty [60, 61]. The accumulation of visible pigmentation in the skin is dependent on the biosynthesis and distribution of melanin [62]. Melanin plays an important role in protecting against UV-induced dermal irritation, but the increased production of cutaneous melanin can cause various skin problems, including melasma, postinflammatory hyperpigmentation, freckles, or lentigines [63]. Therefore, the use of skin-whitening agents has increased tremendously in recent years. During the last decade, skin-whitening cosmetics have become the largest and continuously growing segment of the skin-care market in Asia and have a significant economic value [64].

The melanin pigment is a polymer derived from l-tyrosine and is produced by melanocytes in the human skin. Tyrosinase plays a key role in the regulation of its biosynthesis [65, 66]. As mentioned above, skin pigmentation is influenced by the amount and type of melanin biosynthesized. Therefore, regulating the activity of tyrosinase and the distribution of melanosomes are considered useful strategies for skin depigmentation [67]. Different modalities, including chemical agents and physical therapies, have been introduced to treat hyper-pigmentation and lighten the skin [68]. As described previously [59, 69], most skin-whitening agents target the inhibition of tyrosinase. For example, hydroquinone and kojic acid are the most commonly used skin-whitening compounds, acting as depigmenting agents by inhibiting the tyrosinase enzyme. A number of tyrosinase inhibitors from both natural and synthetic sources have been identified and reviewed elsewhere [65, 70]. However, many tyrosinase inhibitors are inadequately water soluble, which gives rise to skin penetration and formulation problems when these compounds are used directly as skin-whitening agents [70]. To overcome the drawbacks of these active compounds, it is necessary to enhance their targeted delivery to develop efficient cosmetic products.

The use of excipients, such as penetration and solubility enhancers, plays an important role in the absorption of the active ingredients in pharmaceutical and cosmeceutical products [15]. For example, the encapsulation of linoleic acid, a fatty acid with skin-whitening activity, in liposomes significantly enhances its skin-whitening effect because its solubility in aqueous solution is otherwise low [71]. CDs are important and versatile excipients used to enhance the solubility, stability, safety, and bioavailability of bioactive agents [72]. The central cavity of CD is hydrophobic, whereas the rim of the CD's walls is hydrophilic. The hydrophilic external surface and hydrophobic inner surface allow CDs to form more hydrated and stable noncovalent bonds with different guest molecules, altering their physicochemical properties, reducing their undesirable effects, and linking them to various polymers [72, 73]. For example, the encapsulation of linoleic acid in CDs forms soluble complexes and enhances its thermal stability [74, 75]. In the food and cosmetics industries, CDs have been used as complexation agents to increase the water solubility and stability of various chemicals.

CDs are a group of natural cyclic oligosaccharides discovered in 1891. The most common of these molecules are the α-, β- and γ-CDs, composed of six, seven, and eight glucose units, respectively, linked by α-(1, 4) bonds [72]. Because CDs have various advantages in the cosmetics industry, such as improving the solubility and stability of their guest molecules, research efforts have been devoted in the past decade or so to developing new applications for CDs and to understanding their delivery mechanisms [76, 77]. The applications of CDs in cosmetic products were discussed by Buschmann and Schollmeyer [12] in a review in 2002. Empty (uncomplexed) CDs can be used as active agents in cosmetic formulations for several purposes, including the stabilization of emulsions without the use of surface-active agents, the entrapment of unpleasant odors, and the absorption of fatty components [76]. For example, CDs can be used in the formulation of simple O/W and multiple oil–water–oil (O/W/O) emulsions to reduce the need for the high levels of classical surfactants normally required to stabilize emulsions [78]. The main advantage of using CDs to stabilize emulsions is their weak irritant potential compared with those of traditional surfactants, especially hydrophilic-surface-active components [76]. In contrast, CDs may interact with emulsifiers, inducing the coalescence of oil droplets, so using CDs as an excipient can sometimes be difficult [79]. It is also important to consider the potential competition and/or interaction between CDs and other excipients (polymers, cosolvents, emulsifying agents, preservatives) [76].

The most notable feature of CDs is their capacity to form inclusion complexes with hydrophobic guest molecules, including skin-whitening compounds [80]. The encapsulation of skin-whitening agents with CDs does not change the intrinsic bioactive properties of the agents but does change their physicochemical properties [81]. When it forms an inclusion complex, a guest molecule is bound temporarily or entrapped within the inner cavity of the CD host molecule. The main driving force of complex formation is the release of enthalpy-rich water molecules from the cavity [82]. Natural CDs and HP-β-CD are regarded as nonirritant to the skin, eyes, and mucosa upon inhalation [76]. Generally, no type of CDs can permeate the skin barrier in significant amounts, and therefore they do not induce irritation or allergenic reactions [79]. Today, CDs are widely used in conventional formulations, in new cosmetic delivery systems for skin lightening, and in other cosmetic products [76, 83].

The encapsulation of skin-whitening agents by CDs in cosmetics was developed in the late 1980s and has attracted widespread interest from researchers and industry over the past 30 years [12, 84]. It presents many advantages for the application of skin-whitening agents in cosmetic products, in that it enhances the water solubility and chemical stability of the guest molecule, avoids any unwanted adverse effects of the guest molecule, and allows the controlled release of the guest molecule. For example, resveratrol is an efficient pigment-lightening compound that reduces not only tyrosinase activity but also melanogenesis-associated transcription factor (MITF)-expressing melanoma cells [85]. However, the high hydrophobicity of resveratrol and its sensitivity to heat and oxidative enzymes limit its application in food and cosmetic products [73]. However, when it forms inclusion complexes with β-CD or HP-β-CD, the weak solubility of resveratrol can be overcome, and its stability and bioavailability are improved [73, 86].

Cosmetic skin-whitening formulations based on CD inclusion complexes can enhance the whitening effects of the active ingredients and can be applied as milky lotions or in cream dosage forms [12, 87]. For example, hydroquinone is a compound used for skin whitening because it inhibits tyrosinase, thus reducing the conversion of 3,4-dihydroxyphenylalanine to melanin. The oxidation products of hydroquinone, which include quinones and ROS, cause oxidative damage to the tyrosinase protein, resulting in skin lightening [76].

Hydroquinone has been used as a whitening agent for the treatment of hyperpigmented skin disorders for many years. However, in aqueous formulations, hydroquinone is only stable in a limited pH range. Its complexation with CD stabilizes hydroquinone against both oxidation and thermal degradation. Thus, its CD complexes have greater stability because they prevent the oxidation of hydroquinone. They also produce greater depigmentation than hydroquinone itself [12]. Kojic acid is an antibiotic produced by fungal species of the genera *Aspergillus* and *Penicillium* and is also used as a skin-whitening agent in cosmetic products [87], with skin depigmentation activity both *in vitro* and *in vivo*. Similar to other skin depigmenting agents, kojic acid acts as a tyrosinase inhibitor, mainly because it chelates copper. To increase its skin depigmentation efficacy, it is usually used in skin-care products at the highest concentration allowed [68]. However, the use of kojic acid in cosmetics is strictly limited by its light- and heat-labile characteristics. It turns yellowish brown when it decomposes under exposure to light or heat. The CD complexation of kojic acid improves its stability against coloration with time and also improves its skin-whitening effect [88].

Vitamin C (ascorbic acid) is another compound that inhibits melanin formation, by reducing o-quinone formation, and changes oxidized melanin from jet black to a light tan color [68]. However, vitamin C is oxidized rapidly and decomposes in aqueous solution. To overcome these problems, more chemically stable magnesium-l-ascorbyl-2-phosphate (VC-PMG) was synthesized to retain its antioxidant activity and allow its use as a depigmentation agent in skin-care products. However, the formation of CD inclusion complexes is an alternative method for preserving the antioxidant capacity of vitamin C. As previously reported [89], 2-hydroxypropyl-β-CDs can act as "secondary antioxidants", preserving the capacity of a particular antioxidant system produced by CD complexation. Other research has demonstrated that the diffusion of vitamin C is considerably enhanced in the presence of a poly(ethylene glycol)–α-CD complex, and this complex is also useful in enhancing the release of ascorbic acid [90]. Thus, CD inclusion is a promising new technology for developing vitamin-C-containing skin-whitening cosmetic products. The conversion of liquid substances to powder forms is another technological advantage of CD complexation in skin-whitening cosmetics [81]. For example, peroxyacetic acid is a liquid with skin-whitening activity, which is applied as an aqueous solution. However, it forms solid complexes with CDs. These stable powdered complexes are easy to handle and can be used in cosmetic formulations as skin-whitening substances [12].

In the large variety of skin-whitening products available, the use of natural whitening agents is considered desirable and predominates in the cosmetics market [91, 61]. The whitening mechanisms of some natural compounds are not directly related to tyrosinase, but block the upstream regulation of melanogenesis, and these are extremely promising for development of the next generation of whitening agents [91]. Active natural products from plants, such as phenols, flavonoids, coumarins, and other derivatives, have been identified as having skin-whitening properties [69, 92]. Inclusion complexes of natural bioactive products formed with CDs have been used successfully to improve their solubility, chemical stability, and bioavailability [80, 93, 94], which implies a bright future for the development of novel skin-whitening formulations by the encapsulation of natural products. The use of CDs in the design of novel cosmetic delivery systems has also attracted widespread interest in recent years [83]. However, the development of whitening products has focused on identifying new active ingredients rather than on the formulation of delivery systems [64]. Novel CD-based drug carriers, such as self-assembling systems and liposomes, have the advantages of improving the stability and bioavailability of the drugs carried [76, 95], and have a promising future in the delivery of skin-whitening agents and in other cosmetic and dermatological fields.

In summary, the complexation of skin-whitening ingredients with CDs enhances the solubility, increases the stability, and improves the bioavailability of the guest molecule, including synthetic molecules and natural products. The guest molecule is maintained in a more rigid form, its reactive conformations are inhibited, and the molecule is isolated from the environment, reducing any incompatibility with it. The use of CDs in the design of novel delivery systems, such as liposomes, should satisfy the requirements for novel, highly efficient skin-whitening products.

6.4 Molecular inclusion of antiaging agents and its application in cosmetics

As the trend in improving beauty by improving health increases, consumers desire cosmetics with more than basic features, such as comfort and pleasure. They require cosmetics that have added functionality, such as antiaging properties, and most significantly, health and beauty benefits, so that the consumers can live more natural and healthier lives. With the rapid development of new technologies and sciences in recent years, the cosmetics industry has benefitted from techniques such as molecular inclusion [96].

Cosmetics are created to transfer active substances into the human skin when they contact the body. Specific examples are the cosmetics that transfer antiaging substances into the skin. The goal is simply achieved when pharmaceutical and cosmetically active substances are delivered in inclusion compounds so that the natural movement of the body and skin is slowly revitalized and refreshed. The molecular inclusion technology is an alternative method for achieving these desired beneficial effects with satisfactory efficiency [96].

Because it is both flexible and versatile, molecular inclusion is increasing rapidly in the area of cosmetics development. One of the main benefits of the molecular inclusion technology is its capacity to shield active compounds from harmful environmental factors, including heat, moisture, alkalinity, acidity, evaporation, and oxidization. It concurrently prevents the polymerization or degradation of the active ingredient by shielding it from reactions with other compounds in the system. Regulated release, which seems the best option for minimizing environmental damage and increasing efficiency, is another significant benefit of this versatile technique [97, 98].

To construct inclusion complexes, various active compounds are encapsulated in polymers or dispersed as solids in liquids. The encapsulated contents are released under regulated conditions to suit a particular purpose [99, 100]. The mechanism used to release the encapsulated contents depends on the host molecule selected and more significantly, on the exact end use of the product. Factors such as pressure, change in temperature, the pH environment, diffusion, light intensity, biodegradation, the dissolution of the polymer wall, and diffusion through the host molecule can release the core contents of a complex [101, 102]. CDs are currently the most commonly used host molecules in supermolecular chemistry. They can form host–guest inclusion complexes by molecular complexation with several substances to improve the stability and solubility of the guest molecule [103, 104].

The phytoalexin resveratrol occurs naturally and is produced in response to the need for an antiaging agent [105]. It has been discovered in at least 72 plant species, some of which are components of our diets, including grapes, mulberries, and peanuts [106]. Research has shown that resveratrol is effective against free-radical and oxidative damage and prevents the activation and aggregation of platelets [107]. Although the science behind the biological activity of resveratrol is not well understood at the molecular level, resveratrol is known to function as a radical scavenger. However, it is barely watersoluble, which may explain its negligible absorption when administered orally. This shortcoming can be overcome by the creation of inclusion complexes with CDs, which can alter the physiochemical properties, including the stability, solubility, and bioavailability, of poorly water-soluble drugs [108, 109]. However, unmodified or unsubstituted β-CD is unsafe because it is nephrotoxic and is very poorly soluble in water, so several relatively safe and modified CDs have been created, including sulfobutyl ether-β-CD and HP-β-CD. The complexation of *trans*-resveratrol with β-CD has been investigated with reversed-phase liquid chromatography [110]. The phase solubility of resveratrol with HP-β-CD or β-CD shows a linear correlation between the concentration of CD in solution and the amount of resveratrol solubilized. In accordance with a previously reported theory [38], this can be attributed to the creation of 1:1 resveratrol–CD inclusion complexes. As the concentration of CD increases, the solubility of resveratrol also increases in the sequence β-CD < HP-β-CD. The resveratrol–HP-β-CD complex displays increased antioxidant efficacy in terms of its rate and capacity of scavenging the 2,2-diphenyl-1-picrylhydrazyl (DPPH) radical. The complexes created preserved the antioxidant and antiaging activities of resveratrol, evident as the very few differences in the effects of the same concentrations of free and complexed resveratrol.

Phytoalexin is present in most plant stilbenes and is derived from monomeric *trans*-resveratrol. The creation of stilbenes is regarded as part of the plant's defenses, because they display strong antioxidant and antiaging activities [111]. *Trans*-resveratrol occurs in both berries and grapevine tissues, and also in cell cultures, in response to both biotic and abiotic stress. Since the involvement of *trans*-resveratrol in the health benefits conferred by the moderate consumption of red wine was first hypothesized, it has become one of the most widely studied natural products [112]. The cancer-fighting action of *trans*-resveratrol, which prevents carcinogenesis in the three stages of tumor development, is one of its most remarkable antiaging activities [113]. Therefore, as more research data provide fascinating insights into the influence of this compound on the lifespan of various organisms, *trans*-resveratrol can be considered a prospective antiaging agent [96]. Furthermore, because the accumulation of *trans*-resveratrol in culture medium has no harmful effect on cell lines, it allows the formation of successful subcultures.

Cyclodextrins not only induce *trans*-resveratrol biosynthesis but also promote the adducts that eliminate *trans*-resveratrol from cell culture media, reduce the degradation of *trans*-resveratrol and its feedback inhibition, and permit its build-up to high concentrations [114]. Lijavetzky *et al.* [115] previously analyzed the effects of methyljasmonate, CDs, and a combination of both methyljasmonate and CDs on the extracellular production of *trans*-resveratrol. In cells suspension-cultured in methyljasmonate and individually treated with CDs or treated with CDs combined with methyljasmonate, *trans*-resveratrol production decreased as the initial cell density increased. The reduction was more severe in CD-induced cell cultures than in those induced with the combined treatment. The combined use of CDs and methyljasmonate induced the extracellular production of trans-resveratrol.

Linoleic acid is vital to the barrier function of the skin. It is attractive for use in dermatology and cosmetics because of its restorative, antiaging, and skin-care properties [116]. The barrier function of the skin is disrupted when it lacks linoleic acid, and this increases the rate of transepidermal water loss, so the skin becomes dry and turns an unhealthy color. Fatty acids are also vital to the human body. The recommended daily intake for adults is about 8–10 g, and a higher amount is required after serious trauma and during certain diseases [117]. However, the tendency of fatty acids to become rancid and their sensitivity to oxidation restrict the use of polyunsaturated ω-6 fatty acids (PUFAs) in antiaging cosmetic products. In the cosmetics industry, "vitamin F" often refers to a mixture of PUFAs. Vitamin F consists of polyunsaturated ω-6 and ω-3 fatty acids with chains of 18 or more carbon atoms, which contain two or more double bonds. Linoleic acid is the primary ingredient of vitamin F. Linoleic acid is efficiently shielded from oxidation when used as an antiaging agent by formulating it as a molecular inclusion compound with α-CD. Linoleic acid forms inclusion compounds with γ-CD, β-CD, and α-CD, but only the α-CD complex is easily created by the precipitation of the guest and CD in a 1:4 or 1:3 molar ratio. Because the water solubility of the complex is low, the surplus γ-CD, β-CD, and α-CD can be readily eliminated with a washing step. Several studies have been undertaken to determine the appropriate host molecule for linoleic acid, the performance of the inclusion compound, and the optimum complex composition. It was clearly evident from the results that α-CD, which has the smallest cavity, provided the best stabilization effect. A comparison of the different storage stabilities of various CD–linoleic acid complexes confirmed this. Reversible complexation has allowed the use of linoleic acid in several personal-care products and cosmetic formulations for the first time. Therefore, vitamin F, now in a form that is stable to oxygen, light, and temperature, is a very effective molecular inclusion complex for use in cosmetic formulations. These lipophilic vitamin complexes can be regarded as "the world's smallest beauty cases".

For centuries, essential oils have been used for food, traditional medicines, perfumes, and cosmetics [118, 119]. They are complex mixtures of volatile aromatic compounds synthesized in plants to provide protection from several pathogens. Several types of studies have examined the properties and chemical compositions of essential oils. The antibacterial, anticancer, insecticidal, antioxidant, and antiaging activities of essential oils and their components have also been reported [120, 121]. However, their applications are limited by their low water solubility and their volatility. Complexation with CDs is an effective way to improve the aqueous solubility of these substances and to protect them from thermal degradation, evaporation, and oxidation [122]. Moreover, the reversibility of such complexation creates regulated release systems for the encapsulated aroma compounds. The geometric accommodation and polarity of the essential oil molecules and the CD cavity [122] play vital roles in the formation constants and stoichiometries of the inclusion complexes [123]. Monoterpenes, a large group of essential oil constituents, are produced by the condensation of two isoprene units. β-CD is the most frequently used of the various CDs because its cavity size is suitable for a wide range of guest molecules. However, its application is restricted by its low aqueous solubility and its nephrotoxicity [124]. Safer β-CD derivatives with better water solubility have been synthesized, such as HP-β-CD, which is one of the CD derivatives most extensively used in the pharmaceutical, food, and agriculture industries [125]. The inclusion of monoterpenes within HP-β-CD has become popular in recent years [126].

With the exception of thymol, all the monoterpenes studied have shown low or no fluorescence in the absence of HP-β-CD. Thymol emits noticeable fluorescence because it contains an aromatic ring moiety. After the addition of HP-β-CD, the fluorescence intensity of monoterpenes improves and gradually intensifies as the monoterpenes increase. The same increase in fluorescence intensity after complexation with CDs has been demonstrated for several hydrophobic compounds, such as eugenol (EG), anethole (AN), aflatoxins, and acyclovir [127, 128]. The improvement in fluorescence intensity suggests that the monoterpenes are effectively inserted into the CD cavity, creating noncovalent inclusion complexes. The improvement in fluorescence might be caused by several interactions between the encapsulated guest molecule and CD. For example, the spatial restrictions within the cavity could increase the rigidity of the guest molecule [129] and inhibit its "free rotor" effect [130]. The nonbonding electron pairs in the glycosidic oxygen bonds localized within the cavity also confer a high electron density on the hydrophobic cavities of CDs [129]. The inclusion compound shields the fluorophore from external quenchers and reduces its interactions with water [131]. The improvement in fluorescence intensity correlates with a bathochromic shift in the maximum emission wavelength. A hypsochromic shift in the maximum excitation wavelength was also detected (data not shown). These observations are evidence of the translocation of the guest molecule from water into the hydrophobic cavity of HP-β-CD, which is a less polar medium [132].

Improved fluorescence and maximum wavelengths shifts were observed for all the monoterpenes studied after the addition of HP-β-CD, confirming that they were encapsulated within the HP-β-CD cavity. However, to the best of our knowledge, very few studies have determined the formation constant (K_f) values of the HP-β-CD complexes formed with monoterpenes. Saito *et al.* [126] determined the K_f for limonene, linalool, β-pinene, α-pinene, and geraniol inclusion complexes with HP-β-CD, using a constant CD concentration for different guest molecule concentrations. In another study, the guest concentration was constant and the CD concentration was varied [133]. A thermogram of HP-β-CD revealed a very wide endothermic band between 100°C and 150°C, indicating a dehydration process. The differential scanning calorimetry (DSC) curves for β-pinene and linalool showed sharp endothermic peaks at about 204°C and 167°C, respectively. However, the thermal behavior of individual components varied from those of the freeze-dried inclusion complexes. The disappearance of the endothermic peak corresponding to the free-guest molecule suggests complex formation. Similar observations were made for the β-CD–thymol inclusion complex [134]. The broad endothermic peak analogous to free HP-β-CD shifted, indicating that the guest molecules had replaced the water molecules in the HP-β-CD cavity after the formation of the inclusion complex. The same effect, although less intense, was observed in their corresponding physical mixtures. The results of a DSC analysis can be explained by the diffusion of these highly volatile guest molecules into the CD cavity during the heating process [135]. Therefore, the DSC results confirmed that the monoterpenes were effectively included in the cavity of HP-β-CD.

The low water solubility of essential oils might explain why their inclusion rates are lower than those of monoterpene hydrocarbons. Increasing the starting ratios of HP-β-CD to highly hydrophobic guest molecules may be sufficient for their complexation and may produce higher inclusion rates for these molecules [136]. On the contrary, the solubility of the guest molecule in the medium can be improved by the addition of a small amount of water-miscible solvent [136]. Shukla *et al.* [136] found no statistically significant difference between essential oils complexed with HP-β-CD or with β-CD.

Molecular inclusion is a temporary or permanent process in which a material containing an active ingredient (core material) is encapsulated in a shell of a second material. A clathrate compound with a diameter in the nanometer range is called a "nanocapsule" [99, 137]. Today, delivery systems and microcapsules play important roles in the cosmetics and personal-care industries. They offer ideal, unique carrier systems for active compounds, allowing their targeted release, improving their stability and efficacy, isolating and protecting them, allowing their safe administration, masking their undesirable properties, and improving the tactile and visual appearance of anti-aging cosmetic products.

Regulated release technologies are used to deliver the prescribed amounts of antiaging agent with improved efficacy, convenience, and safety. In these industries, there is a constant need for new delivery systems for the safe incorporation of the new and sensitive active ingredients of novel antiaging formulations. Innovations in the area of cosmetic delivery systems should lead to simpler and easier systems and improve the development of important emulsion systems. Sensitive antiaging ingredients must usually be added to these systems in specific and complex ways, under highly controlled conditions (e.g., temperature and water/oil content). Molecular inclusion can be used to deliver anti-aging ingredients to some difficult systems, such as those containing α-hydroxy acids, salicylic acid, those with high alcohol contents or glycolic acid, and critical water-in-oil (W/O) or silicone emulsions. They can be used to transfer active antiaging ingredients into the skin in an effective, targeted, and pain-free manner; shield compounds; shield fragrances or volatile compounds from evaporation; shield active ingredients from degradation by heat, moisture, or light; and regulate the rate of their release [138, 139]. Molecular inclusion can be used in antiaging cosmetic formulations, such as bath gels, makeup, tanning creams, lotions, perfumes, hair products, and soaps. Molecular inclusion can improve the personal-care and cosmetic industries because it is innovative and allows the production of valuable antiaging products, in response to human needs and desires [140]. A delivery system can be defined as a method of holding, carrying, and transporting an antiaging ingredient. It refers to any type of vehicle that delivers an active ingredient to a target site. Molecular inclusion systems can also regulate the release rate of an active ingredient from a cosmetic product to the ideal level. Several pathways can transport antiaging agents through the skin. The intercellular route occurs at the boundaries between cells, through the lipid bilayers, in a tortuous permeation pathway. In contrast, transcellular pathways can pass agents directly through cells. Transportation is also possible through hair follicles or sweat ducts [141, 142].

The topical application of cosmetic antiaging products often requires the effective delivery of active ingredients through the skin's barriers to reach the target skin layer. A layer of cells joins the epidermis to the stratum corneumand presents an obstacle to effective transdermal delivery because it is highly impermeable. The inner epidermal and dermal layers of the skin are significantly more hydrophilic, containing 50% and 70% water, respectively, whereas the lipophilic stratum corneum contains only about 13% water. Various factors, such as age, the external environment, and the varying permeability of the skin at different body sites (e.g., face vs. legs vs. palms) together influence the skin's barrier function [143–145]. Molecules that can pass through the epidermis are reportedly limited to those of low molecular mass (5500 Da) and moderate lipophilicity. They include molecules that are sufficiently soluble in the lipid domain of the stratum corneum, while still sufficiently hydrophilic to allow their partitioning to the inner layers of the skin. Molecular inclusion techniques, with the appropriate shell materials, are essential because some cosmetic antiaging substances are too hydrophilic to pass through the stratum corneum or too lipophilic to partition to the epidermis and require a system to provide the level of lipophilicity required for specific applications. Mathematical models based on quantitative structure–permeability relationships, diffusion mechanisms, or a combination of both can be used to predict skin permeability. At the same time, a compound must have the hydrophilic or lipophilic characteristics that make it soluble in the cosmetic antiaging formulation itself and ensure its stability during the formulation, storage, and application of the product [142, 143, 146].

Several important aspects must be considered when cosmetic antiaging active substances are delivered through the skin. These include the right concentrations of the components, the correct application time of the product to the skin, and the correct site of action of the cosmetic ingredient. Other factors include formulation type, droplet size on the skin, polarity of the formulation, polarity of the stratum corneum, and the organization of the skin lipids. Molecular inclusion and its delivery system must retain the functional molecule in a specific skin layer and avoid undesirable transdermal delivery [142, 147].

The correct choice of the outer shell material is very important, and must be consistent with the proposed application, because it affects the stability and efficiency of the inclusion compound. Factors to be considered when selecting a wall material for topical applications include its viscosity, mechanical properties, toxicity, stability, biocompatibility, compatibility between the active ingredient and the host molecule, improvement of the active penetration to the stratum corneum, the anticipated particle size, the release of the antiaging ingredient from the transport vehicle into the skin, the microscopic properties of the surfaces of the microparticles, and economic and processing factors. Mixtures of inclusion compounds frequently achieve the desired outcomes because most individual encapsulating materials do not have all the necessary properties [148]. An inclusion compound can be chosen from several different synthetic and natural polymers. Shielding the active compounds of antiaging cosmetics in microsized carriers, to be released in a controlled way over a definite period of time, has been a topic of significant research. The host molecules most frequently used in cosmetic antiaging formulations include proteins, polysaccharides, lipids, and synthetic polymers. Inorganic substances, such as clays, polyphosphates, and silicates, can also be used as secondary polymers [149–151]. Their properties are well defined by their structural characteristics, such as their molecular weight, the composition of the copolymer, and the nature of the chain endgroups, and these polymers can be chemically modified to confer enhanced functional properties [143, 152–154].

6.5 Molecular inclusion of moisturizing agents and its application in cosmetics

When using the molecular inclusion method to deliver a cosmetic moisturizing agent to the skin, several important qualities must be considered. Product quality and economic, technical, and ethical issues must also be considered in the large-scale manufacture of cosmetics that use molecular inclusion and its delivery systems. The process of scalingup a method to achieve greater production is a major difficulty faced by industry [155]. Although complexed compounds can sometimes be used directly in products, they must usually be attached to a substrate or suspended in a medium to function properly. Two drawbacks often faced are finding the best process for attaching the moisturizing agent to the host molecule and the prevention of undesirable effects of the substrate on the stability of the complexed compounds. Undesirable effects on cosmetics containing moisturizing agents that affect the stability of the complexed compounds must be avoided. These include the use of a moisturizing agent that encroaches into the host molecule before or during the application period or leaches the moisturizing agent from the complexation compound or include factors that hinder the successful release of the complexed compound. Cosmetic moisturizing agent preparations should be constructed in such a way that the internal cavities and the outer surfaces of the host molecules remain intact throughout the formulation period, the transport period, and the storage period before use [156]. Complexed compounds used in cosmetics that contain moisturizing agents should be effortlessly squashed on the skin when applied, to release the core components, and should also be sufficiently rigid that their shape is not altered during the manufacture, storage, or transport of the cosmetic preparation because enough moisturizing agent must remain encapsulated at the desired release time [157]. Products containing complexed compounds are stored in different environments. Elements that must be considered include humidity, temperature, pressure, light, other forms of radiation, and air pollutants. However, some moisturizing agents with unstable or highly reactive components must be shielded from extreme temperatures to prevent their premature decomposition or the evaporation of the guest molecule. Extreme stress on the complexed compounds during storage, such as that occurring in some slow-release fertilizers, may cause blockage or fusion so that the product is no longer free to leach through the surface to which it is applied.

When the moisturizing agents in cosmetic formulations are biodegradable substances, both bacterial and fungal degradation can be a problem. At times, it is important to add preservatives to increase their stability, but the adverse effects of these preservatives must be monitored and limited. Grapefruit seed oil and tea tree oil are effective natural remedies. Benzalkonium chloride added to them in small amounts acts as an emulsifier, an antimicrobial agent, and a moisturizing agent [152]. Cosmetic moisturizing agents and delivery techniques quickly come to and go from the market, and it is important that they are based on technological advances that consumers expect. As new molecular inclusion techniques emerge, they should be adapted rapidly to new active compounds in ways that are both simple and dynamic. When choosing an appropriate guest molecule and delivery system to improve the stability, aesthetic appeal, efficacy, and safety of the final product, it is important to consider the economic and technical aspects of products. Understanding the willingness of consumers to pay the additional cost associated with complexed compounds and correctly judging the general economic parameters of a moisturizing product are probably the most important criteria. The cost of production, consumer education, marketing, price, and distribution must all be considered. The only way a product becomes successful is through its appeal to the consumer, its user-friendliness, and the willingness of consumers to pay for it. Sadly, there is no one solution to this computation, so meticulous market studies and product development are required [83].

The safety of the product is also important because it ensures that the product is sustainable and purchased repeatedly. Information on the composition of the product and any traces of hazardous materials, such as heavy metals, must be accessible to consumers. The product must be deemed toxicologically safe by the appropriate consumer protection agencies and perceived as safe by consumers at large. The manufacturing process, the complexation compound, and the distribution and safety of a product must comply with the laws and regulations governing the cosmetics industry in the country into which it is to be introduced. The dermatological and marketing perspectives, together with the safety and legal requirements of the skin delivery systems and cosmetics, must all be considered. European cosmetic regulation 1223/2009 must be carefully considered because it presents new rules regarding the use of complexed compounds in cosmetic products [158].

The ability of CDs to form solid inclusion complexes (host–guest complexes) by molecular complexation with various liquid, gaseous, and solid compounds is their most distinctive feature [159]. The guest molecule occupies the cavity within the CD molecule in these complexes, which occurs when the dimensions of the guest molecule are congruent with those of the host cavity [160]. Inclusion complexes form when adequately sized nonpolar moieties enter the lipophilic cavities of CD molecules [7]. The formation of inclusion complexes does not involve the formation or breakage of covalent bonds. The release of enthalpy-rich water molecules from the cavity is the driving force behind the formation of the complex. These molecules are displaced by the guest molecule, which is more hydrophobic and available in the solution, forming an apolar–apolar association and reducing the energy state by reducing the ring strain on CD and stabilizing it [161]. A dynamic equilibrium exists between the binding of the moisturizing agent within the host CD. Many different guest molecules that also act as moisturizing agents are available for molecular inclusion in CDs, including guest compounds. The availability of several reactive hydroxyl groups allows the functionality of CDs to be increased by chemical changes, which expand their applications. The substitution of different functional groups on the primary and/or secondary faces of the CD molecule modifies it. The substituted functional groups of these modified CDs can act in molecular recognition and are very useful in complexation. Analytical chemistry and the targeted delivery of moisturizing agents utilize this same property because modified CDs have better enantioselectivity than native CDs [82].

Urea is commonly used in skin-care and dermatological formulations because it is a natural moisturizer of the stratum corneum. It is also a penetrating moisturizer because it has a very strong osmotic effect. However, a major disadvantage of urea is its instability in products containing water, in which it becomes biologically unavailable and can cause discolorization and emulsion instability when it decomposes. Urea also crystallizes on the skin surface when it evaporates. Acidic urea products can cause stinging or burning sensations, but urea products must be kept at an acidic pH to prevent the decomposition of urea to ammonia. Haddadi *et al.* [143] formulated and characterized O/W and W/O creams consisting of urea-loaded microcapsules used to stabilize the urea. The pattern of urea release differed among all the formulations examined, and the release from the W/O creams showed zero-order kinetics, whereas the release from the O/W creams followed Higuchi kinetics. The release from the creams containing microparticulate urea was slower than from the creams containing free urea.

Several cosmetic products contain glycolic acid, which functions as a moisturizer. Unfortunately, the probability of skin irritation and burning increases as the benefits of the glycolic acid cosmetic increase. Therefore, topical controlled-delivery systems loaded with glycolic acid may enhance the cosmetic properties of the acid while reducing its adverse effects. Perugini *et al.* [162] assessed several kinds of molecular inclusion systems for the encapsulation of glycolic acid, including liposomes, chitosan microspheres, and liposomes modified by the addition of chitosan. *In vitro* trials were performed to evaluate the viability of the complexed compounds in controlling the release of glycolic acid. The results indicated that liposomes are most appropriate for modulating the release of glycolic acid and that the best conditions for achieving this control are liposomal systems with a glycolic acid:lipid molar ratio of 5:1. The addition of chitosan to the liposomes substantially increased the control of release, whereas chitosan microspheres did not allow controlled glycolic acid release, even after they were cross-linked.

Although the present delivery systems for the transdermal application of cosmetic formulations appear promising, the molecular inclusion methods discussed above must still be clarified. At present, only the primary scientific and technical requirements for molecular inclusion have been addressed. More research is necessary to better understand the modifications of these materials, their effects, and transitions at every stage from the manufacture of the product to its final form, how they affect the penetration of the product, and how they modify the diffusion of the active substances into the skin [83]. Although transdermal delivery is relatively well understood, "real-life" data must be collected in future studies because the main factors affecting dermal delivery are still unknown. It is difficult to measure this domain, and numerous regulations and parameters must be considered throughout the formulation process. Several methodologies for molecular inclusion techniques are available only in published patents and are yet to be applied on an industrial scale [142].

Nonetheless, the relevant clinical trials of these cosmetic products have shown the tremendous potential of these molecular inclusion systems, which provide attractive results. There is great interest in how the delivery systems affect the speed required to achieve the maximum efficacy or intensity of the cosmetic effect of the formulations. With the very aggressive regulations in mind, it is imperative to use approved complexation compounds to create new properties. Today, cosmetic products are more about personal care than merely beauty, because they stimulate and balance the skin functions, and repair, hydrate, regenerate, and vitalize the skin. Many also successfully complement several treatments for skin disorders.

With the increasing use and availability of so-called "cosmeceuticals", the gap between cosmetics and pharmaceuticals is rapidly closing. Such products contain many natural substances, including therapeutic extracts, vitamins, and moisturizing agents, which greatly benefit from the molecular inclusion technology. In the future, molecular inclusion will be used as an important tool for targeting specific cells with better delivery (e.g., targeting adipocytes to reduce fat storage or melanocytes to reduce pigmentation mechanisms is crucial to increasing the activities of these cosmetic formulations), for shielding unstable ingredients, for reducing or eliminating the adverse effects of molecules by improving their efficacy, and for dosage reduction to improve marketing. Consumers are likely to approve systems that release various moisturizing agents in response to temperature or pH modulation in the future [163]. These delivery systems offer tremendous hope for the future of the cosmetics industry.

6.6 Molecular inclusion of perfume agents and its application to cosmetics

The microencapsulation of flavors and fragrances is a topic of both academic and industrial interest, as shown, for example, by the numerous scientific symposia devoted to the subject. Fragrant chemicals are added to a large variety of consumer products, including perfumes, laundry detergents, fabric softeners, soaps, detergents, and personal-care products, and aroma chemicals are widely used by the food industry to enhance taste. Therefore, the flavor and fragrance industry is an important sector of the chemical industry (sales were estimated to be $21.8 billion in 2011) and is traditionally an important contributor to research in organic chemistry. Today, this industry uses an extensive range of encapsulation technologies, because microencapsulated flavors and fragrances improve the efficacy demanded for a wide range of applications, including cosmetics (perfumes), personal care (hand and body wash, toothpaste, etc.), food (flavors), and home care (laundry detergents). In cosmetic, household, and personal-care products, a common objective is to extend the life and improve the delivery of highly volatile fragrances, with the gradual liberation of the encapsulated functional species, i.e., the controlled release of the active ingredient. There is also great interest in infusing scents into everyday materials, such as fabrics, or in turning liquid flavors into free-flowing powders to increase their shelf lives in food products. Significant health and environmental benefits may also be derived from the efficient encapsulation of fragrances. For example, most of the synthetic nitro- and polycyclic musks used in perfumes, deodorants, and detergents are toxic and nonbiodegradable. As a result, they tend to accumulate in the environment and in human milk. Natural fragrances that have not been used because their chemical and physical stabilities are poor might replace nonbiodegradable musks when the labile fragrance molecules are chemically and physically stabilized in a sol–gel silica matrix.

CDs can be complexed with the different fragrance molecules included in personal-care products, such as shampoos, deodorants, detergents, and absorbent powders, including bath and baby powder products. CDs may be useful in masking the unfavorable organoleptic characteristics of some cosmetic products, attributable to the presence of a particular active agent. For example, dihydroxyacetone is used as a tanning agent and is characterized by an unpleasant odor that is difficult to cover with perfumes. However, when a dihydroxyacetone–CD inclusion complex is used, this odor vanishes completely. Furthermore, complexation with CD allows this agent to be released in a gradual manner, ensuring uniform tanning of the skin. Glutathione, which is used as a depigmenting and whitening agent, also confers an unpleasant smell on cosmetic products and has been successfully complexed with different CDs. CDs are also used to cover the odors of the mercaptans used in permanent wave products, thus avoiding the use of perfumes, which may induce allergic reactions.

Several authors have demonstrated that CDs, and in particular β-CD, can interact and form complexes with the different components of sweat and body secretions that are responsible for body odor, such as thiols, steroids, and acids. For this reason, β-CD could be used as an active component of deodorants and other personal-care formulations. Other authors have investigated several unsaturated aldehydes, such as hexenal, octenal, and nonenal, that are responsible for the so-called "aging odor", because their levels are markedly increased in the middle-aged and elderly. These researchers have demonstrated that M-β-CD is the most effective CD in the deodorization of "aging odor".

The most familiar smell of humans is our personal odor, generated by our own body. Body odor can be classified into two types: (i) that originating from the whole body, including sweat glands and the skin surface; and (ii) that originating from a specific part of the body, such as foul breath, underarm odor, odor from the skin or hair of the head, and foot odor. In many cases, body odor is perceived as a combination of several odors, whereas those emanating from a specific part of the body tend to be characteristic of that body part. Common contributors to body odor are the secretions of the skin's sebaceous and sweat glands and the waste-product plaque from the stratum corneum. The metabolism of these compounds by the bacterial flora on the skin or by atmospheric oxidation then produces a volatile odor. The human scent is known to vary across individuals in both its strength and quality and can be considered an important signal of a person's state of cleanliness and overall physical condition. Various factors, including sex, age, eating habits, living environment, and race, can all affect body odor. Although the chemical composition of the body's odor includes mainly low-molecular-weight volatile fatty acids, sulfur compounds, and steroids, these may differ in specific parts of the body and under certain physical conditions.

Body odor is usually associated with a negative impression, so humans are normally keen to eliminate or markedly reduce it. Body odor inhibits social interactions by reducing self-confidence. Consequently, antiperspirants, deodorants, and herbal products have been used extensively to treat body odor. However, some effects limit the use of different approaches, such as the application of high levels of fragrance to mask malodor or changing the balance of the skin's bacterial population. CDs are a new class of cosmetic ingredient that can be used to control odor by forming complexes with the molecules responsible for an unpleasant smell. CDs are an attractive ingredient for encapsulating molecules because they are naturally occurring cyclo-oligosaccharides, characterized by a high degree of safety when used topically. They have an internal hydrophobic cavity in which different lipophilic compounds can be included, whereas their hydrophilic external surfaces usually make the host–guest inclusion complexes water soluble. The physical, chemical, and biological properties of the guest compounds can be altered considerably by their complexation with CDs. Therefore, CD inclusion complexes are widely used for practical applications in the cosmetics industry.

Cyclodextrin-based preparations are used in cosmetic products to reduce body odors. Dry CD powders, with particles of less than 12 μm, are used for odor control in diapers, menstrual products, paper towels, etc. and are also used in hair-care preparations to reduce volatile malodorous mercaptans. Dishwashing and laundry detergents complexed with CDs can mask odors in the washed items. Complexation with the CDs used in silica-based toothpastes increases the availability of triclosan, improving its availability almost threefold.

Dihydroxyacetone is used as a tanning agent but is unstable in aqueous solution. More important is the unpleasant odor of dihydroxyacetone, which is difficult to mask with perfumes. However, this odor vanishes when dihydroxyacetone is complexed with CD. The slow release of dihydroxyacetone from the complex also results in more uniform tanning of the skin. This CD complex is used in Ultrasun Selftan© (Ultrasun) and Self-Action SuperTan For Face© (Estee Lauder).

Glutathione has various physiological activities, including the inhibition of melanin synthesis. Therefore, it can be used for skin whitening and for skin-improving effects. Unfortunately, glutathione generates an offensive odor when used in cosmetic formulations. However, glutathione complexed with CD has no odor of glutathione, while maintaining the same effect on the skin as glutathione.

Mercapto compounds are often used in waving lotions, but the presence of mercapto derivatives generates an extremely unpleasant odor during their application. This odor is often masked, in part, by the addition of perfumes to the lotion. A more efficient solution is the use of CDs, because the unpleasant odor is eliminated when mercapto compounds are complexed with CDs.

Chamomile extracts and oil have antiphlogistic, bacteriostatic, and wound-healing effects. However, these formulations often have an intense unpleasant odor. This odor is reduced by complexation with CDs, whereas the anti-inflammatory activities of the chamomile extracts remain unaffected.

Deodorants are used to control the odors formed by the microbial degradation of sweat. CDs and mixtures of CDs can be used in deodorant sticks because they can form complexes with the molecules that contribute to perspiration odor. Another area of CD use in cosmetic preparations is in suppressing the volatility of perfumes, room fresheners, and detergents by controlling the release of the fragrances from the inclusion compound. The interaction between the guest molecule and CD produces a high-energy barrier to volatilization, thus producing long-lasting fragrances. The major benefits of CDs in this sector are stabilization, odor control, processing improvement when a liquid ingredient is converted to a solid form, flavor protection and flavor delivery in lipsticks, water solubility, and the enhanced thermal stability of oils. Some other applications include their use in toothpastes, skin creams, liquid and solid fabric softeners, paper towels, tissues, and underarm shields.

The use of CD-complexed fragrances in skin preparations, such as talcum powder, stabilizes the fragrance against loss by evaporation or oxidation over a long period. The fragrance is enclosed within the CD, and the resulting inclusion compound is complexed with calcium phosphate to stabilize the fragrance during the manufacture of bathing preparations [164]. Thus, researchers have prepared cosmetic formulations containing CDs to create long-lasting fragrances.

The complexation of some scents, such as rose and lemon oil, with CDs greatly reduces the eye irritation caused by shampoos, whereas in most absorbent powders, CDs can be substituted for starch, circumventing the risk of bacterial contamination; the complexation with CDs also promotes a long-lasting fragrance effect (more than 6 weeks after opening). When they are included in detergents for the complexation of fragrances molecules, CDs also act as antifoaming agents, considerably reducing the water consumed for rinsing. Table 6.2 lists some examples of CDs used for fragrance or aroma encapsulation in different cosmetic applications

Besides Table 6.2, Higueras *et al.* [181] developed films made from chitosan, HP-β-CDs, and glycerol that could modulate the loading capacity and release velocity of carvacrol. They first generated the films by casting and conditioning the mixture at various relative humidities (RHs) and then immersed the films in liquid carvacrol. The addition of CDs to the chitosan system improved the encapsulation of carvacrol. A mixture of glycerol and water was used as a plasticizer to control the loading efficiency of carvacrol into the film. There was a good positive correlation between carvacrol retention and film plasticization. The diffusion coefficient of carvacrol varied greatly with different plasticization and conditioning parameters. RH was a key factor affecting the release of carvacrol. These films also had antimicrobial activity against *Escherichia coli* and *Staphylococcus aureus* at 25°C and 43% environmental RH after 20 days in storage. The films could be used in the design of facial masks for delivering bioactive fragrance compounds in the cosmetics industry.

Mascheroni *et al.* [182] used a single-step electrospinning process to prepare nanofibrous membranes from a blend of edible carbohydrate polymers (pullulan and β-CD). They then used these nanofibrous membranes to encapsulate bioactive aroma compounds and studied their humidity-triggered release. Encapsulation was rapid and efficient, and further studies indicated that the membranes accommodated small, homogeneously dispersed crystals of CD–aroma complexes, which were formed during electrospinning. The release of the aroma compound was very slow under ambient conditions (23°C and 55% RH) and even at high temperatures (up to 230°C), but release occurred above a critical specific relative humidity point (90%), which was ascribed to this type of structure. These nanofibrous

Table 6.2.   CDs used for fragrance or aroma encapsulation in different cosmetic applications.

CD type	Fragrance/aroma/malodor	Aim of this study	Methodology	Results and discussion	Ref.
2-HP-β-CD	Linalool, citronellol, citral, limonene, methyl ionone, benzyl acetate, linalyl acetate, methyl dihydrojasmonate	To dissolve water-insoluble perfumes in cosmetics using 2HP-β-CD without the need for a surfactant	94g water → mixture → stirrer → filtration → Complex solution; 5g 2HP-β-CD; Gradual addition; 1g fragrance	Perfumes with an OH group and an aromatic ester group and low Mw (150–160) were highly soluble in 2-HP-β-CD.	[165]
α-, β-CDs and CD polymers	Linalool and camphor	To investigate the feasibility of preparing novel controlled-release systems for the delivery of essential oils used as ambient odors	15g water → mixture → stirrer → 4h at 60°C → precipitation; 5g NaOH; β-CD; Epichlorohydrin	α-CDs have a greater formation constant and retention ability in aqueous phase for the two guests.	[166]
β-CD and 2-HP-β-CD	Linalool and benzyl acetate	To increase the stability and watersolubility of fragrance molecules, ensure controlled release of these compounds, and improve the handling properties of fragrance materials in cosmetic formulations	47g water → mixture → stirrer → 12h at 25°C → Complex solution; 2.5g HP-β-CD 0.28g linalool and 0.272 g benzyl acetate; Gradual addition	2-HP-α-CD significantly increased the water solubility of linalool and benzyl acetate and handling properties.	[167]

HP-β-CD	Osmanthus flower fragrance	To analyze the release model of a fragrance–HP-β-CD inclusion complex and microcapsules	Vacuum drying at 60°C 1.5h Inclusion complex	The fragrance in microcapsules is smaller than that of inclusion complex.	[168]
CD-intercalated layered double hydroxides	Toluene, limonene, menthone	To investigate the inclusion and the release of three volatile organic guest molecules (two fragrances and toluene as a model compound) in the aqueous and gaseous phases	LDH with CMCD followed by inclusion of guests	Two CMCDs have been intercalated in Mg–Al LDH followed by the inclusion of volatile organic compounds. CMCD-LDH allows the controlled release of aroma compounds.	[169]

(Continued)

Table 6.2.   (*Continued*)

CD type	Fragrance/ aroma/malodor	Aim of this study	Methodology	Results and discussion	Ref.
HP-*β*-CD	Methyl anthranilate, menthol	To observe the release of fluorescent fragrance molecules, such as MA, and nonfluorescent molecules, such as menthol, with fluorescent probes (typically 7-amino-4-methylcoumarin: AMC)		The moisture-activated odor release from fragrance/cyclodextrin inclusion complexes could be visualized by using a fluorescence imaging system.	[170]
HP-*β*-CD	Monoterpenes	To investigate the inclusion complexation of 2-HP-*β*-CD with eight monoterpenes in aqueous solution and the solid state		The encapsulation with HP-*α*-CD could be a very effective vector to design new monoterpenes formulations in cosmetic industries.	[171]

β-CD	(−)-Linalool	To produce β-CD inclusion complexes of (−)-linalool using physical mixture, slurry, and paste methods in stoichiometric ratios	3.783mL water→mixture→stirrer →400 rpm 36h at 25±2°C → desiccator → Gradual addition → Dry pastes/slurry; 1.135mg β-CD 154mg (-)-linalool (1:1 molar guest/host ratio)	(−)-linalool/β-CD inclusion complex could be successfully produced according to the paste or slurry method.	[172]
β-CD	Terpinen-4-ol	To investigate the release performance and antibacterial activity of the β-CD–terpinen-4-ol complex	25mL water→mixture→stirrer →400 rpm 2h at 50±2°C → Centrifuged and freeze-drying → Gradual addition → Solid state; 1.135gβ-CD 0.154g terpinen-4-ol (dissolved in 5 ml ethanol)	Such a solid terpinen-4-ol/α-CD inclusion complex can be quite applicable in cosmetic industries.	[173]
β-CD	Linalool	To prepare and characterize linalool-β-CD inclusion complexes with ESI-MS, DSC, and TG	Linalool + β-CD 1:1 molar ratio → Coprecipitation → filtration → Oven Drying at 50C 24h	The aqueous solubility of linalool was significantly increased by inclusion in α-CD.	[174]
β-CD	Vanillin	To study self- and non-self-aggregation (vanillin–β-CD complexation) quantitatively and to structurally characterize these aggregates	vanillin + β-CD 15 mM → Coprecipitation → filtration → lyophilization	The identification and characterization of phenomena such as self-aggregation and host-guest complexations require "soft" analytical technique.	[175]

(Continued)

Table 6.2.   (*Continued*)

CD type	Fragrance/ aroma/malodor	Aim of this study	Methodology	Results and discussion	Ref.
β-CD	Phenylethanol	To prepare and characterize inclusion complexes of the two phenylethanol isomers with β-CD	25mL water, 1.135g β-CD, 0.122g 1-PE or 2-PE (dissolved in 5mL ethanol) → Gradual addition → mixture → stirrer → 400 rpm 2 h at 50°C → Centrifged and freeze-drying → Solid state	1-PE and 2-PE can form inclusion complexes with ββ-CD in the solid state and greatly enhance their stability.	[176]
HP-β-CD in liposomes (DCL)	EG in clove oil	To develop a membrane contactor to prepare inclusion complexes of EG and CEO and then encapsulate them in liposomes	20mL water, 500mg HP-β-CD, Eug or CEO (1:1 molar ratio) → Gradual addition → mixture → stirrer → 150 rpm 24 h at 25±2°C → filtration and stored at 4°C → Inclusion complex	DCL may constitute a valuable and promising carrier system to protect volatile and poorly water-soluble molecules.	[177]
β-CD	2-Nonanone (2-NN)	To prepare a β-CD–2-nonanone complex and test its antifungal activity against *Botrytis cinerea*	50mL ethanol/water(1:2), 5g β-CD, 2-NN (1:0.5,1:1,1:2 molar ratio) → Gradual addition → mixture → stirrer → 150 rpm 4 h at 35±2°C filtration and convection at 50°C 24h → Solid state	2-NN was complexed with β-CD to form an inclusion complex with the most efficient system by the co-precipitation method in the molar ratio of 1:0.5.	[178]

β-CD	Clove oil (CO)	To optimize a method for preparing solid essential CO–β-CD complexes		The use of microwave irradiation could be an alternative for the aroma industry for preparing soluble and insoluble essential CO–β-CDs complexes.	[179]
HP-β-CD, M-β-CD, HP-γ-CD	Geraniol	To form inclusion complexes of geraniol with three modified CDs and produce a free-standing CD–geraniolinclusion complex nanofibrous web with electrospinning		Electrospun CD–geraniolinclusion complex nanofibrous webs have shown enhanced shelf-life of geraniol along with antibacterial and antioxidant properties, which may have potentials to be used as prolonged releasing systems for various applications in cosmetics industry.	[180]

polysaccharide membranes are sustainable and safe for application in active facial masks.

Baránková *et al.* [183] used an effective headspace gas chromatography (GC) method to study the effects of macrocyclic polysaccharides (α-CD, β-CD) on the air–water partition coefficient of two isomeric esters, ethyl butanoate and butyl ethanoate. They found that inclusion complexation with CDs had the strongest effect on aroma retention. In all cases, these effects correlated well with the concentration of the additive, which can be explained by the 1:1 complex formation model. Thermodynamic studies have shown that ester–CD binding is enthalpy driven. It is useful to explain the different binding affinities with simple molecular modeling. The individual ester–CD pairs are promoted by the solute–CD cavity size match. This study could be used to guide cosmetic technicians in selecting suitable guest fragrance molecules to be hosted by CDs.

Ciobanu *et al.* [184] investigated the complexation efficiency of volatile flavor compounds and CDs with static headspace GC (SH-GC). They selected α-CD, β-CD, γ-CD, HP-β-CD, RM-β-CD, and a low-methylated β-CD (CRYSMEB) to encapsulate 13 volatile flavor compounds (α-pinene, β-pinene, camphene, eucalyptol, limonene, linalool, p-cymene, myrcene, menthone, menthol, trans-AN, pulegone, and camphor). For all these volatile compounds, 1:1 inclusion complexes were formed. The β-CDs usually gave higher stability constants than α-CDs or γ-CDs. Furthermore, the complexation efficiency of native β-CD was almost equal to those of the modified β-CDs (HP-β-CD, RAMEB, and CRYSMEB). This report might help cosmetic scientists to find the best-matched pairs between CDs and volatile aroma compounds.

Estragole is a major component of the essential oils of basil and tarragon. To identify the most suitable host for estragole, Kfoury *et al.* [185] characterized inclusion complexes between a series of CDs (including α-CD, β-CD, HP-β-CD, RAMEB, CRYSMEB, and γ-CD) and estragole. They used SH-GC and UV–visible spectroscopy to determine the formation constants (y) of the complexes with nonlinear regression. They then prepared freeze-dried inclusion complexes with different CD–estragole molar ratios and characterized them with DSC and Fourier transform infrared spectroscopy (FTIR). The controlled release of estragole was attributed to the formation of the inclusion complexes. Inclusion complexes between CDs and estragole or estragole-containing essential oils increased the radical (DPPH)-scavenging activity and photostability of the guest molecules. From these findings, it was concluded that inclusion within CDs is an effective way to improve the application of estragole and estragole-containing essential oils in the fields of aromatherapy and cosmetics.

To achieve high thermal stability and the slow release of EG, Kayaci *et al.* [186] prepared nanofibers from polyvinyl alcohol (PVA) and EG–CD inclusion complexes (EG–CD) with an electrospinning technique. They selected three types of native cyclodextrins (α-CD, β-CD, and γ-CD) to determine the most favorable CD type for the stabilization of EG. They found that α-CD was not suitable for EG inclusion in the case of PVA–EG–α-CD nanofibers. However, the durability and thermal stability for EG were significantly improved in both the PVA–EG–β-CD and PVA–EG–γ-CD nanofibers. The electrospun nanofibers containing CD inclusion complexes of active fragrance compounds, such as EG, may be very useful in the cosmetics industry. This technology has several other advantages, including the extremely large surface area of the nanofibers, their specific functionality, high thermal stability, and the slow release of the active compound from the CD inclusion complex.

Xiao *et al.* [187] prepared an inclusion complex of citral with monochlorotriazine-β-CD (MCT-β-CD) and grafted it onto cotton fabric. They characterized the inclusion complex with FTIR and used Diffraction Laser Spectrometry (DLS) to measure the particle size distribution of citral–MCT-β-CD. They determined the encapsulation efficiency of the inclusion complex with UV–visible absorption spectroscopy and a thermogravimetric analysis (TGA). Finally, they observed the morphology of the fabric grafted with the citral–MCT-β-CD inclusion complex with a scanning electron microscope (SEM). The particle size of citral–MCT-β-CD was approximately 200–300 nm and the encapsulation efficiency was about 9%. The citral release rate from the inclusion-complex-grafted fabric was markedly lower than that from the control. This study may prompt the production of aromatic cotton fabrics, with controlled-release fragrances. This fabric may also be considered for use in facial masks.

Kfoury *et al.* [188] studied the inclusion complexes of *trans*-AN with α-CD, β-CD, HP-β-CD, RAMEB, and CRYSMEB. They investigated their complexation behaviors in aqueous solution using SH-GC, a phase solubility study, UV–visible spectroscopy, ^{1}H nuclear magnetic resonance (NMR), and (two-dimensional) rotating-frame Overhauser effect spectroscopy (ROESY)NMR. All the CDs tested formed 1:1 inclusion complexes with AN. Surprisingly, phase solubility and retention studies showed that the watersolubility of AN was markedly enhanced by its complexation with CDs. Freeze-dried inclusion complexes were prepared, and the encapsulation of AN was confirmed with FTIR and DSC. The degradation of AN, triggered by UVC irradiation, was also significantly reduced by the formation of the CD inclusion complexes. Therefore, inclusion within CDs is an effective way to increase the solubility and stability of AN, making it valuable for cosmetic or pharmaceutical applications.

Thymol is an effective antibacterial agent that can be used as a preservative. Unfortunately, it has low aqueous solubility and a strong bitter/irritating taste. Nieddu *et al.* [189] developed a copolymer made from β-CD and dimethylaminoethyl methacrylate (DMAEMA) to encapsulate thymol to improve its powderization, solubilization, and taste-masking properties. They used coprecipitation and sealed-heating methods to prepare the thymol–β-CD complex and mixed the complex with the DMAEMA copolymer to reduce the volatility of thymol before *in vivo* and *in vitro* studies. Sealed heating was shown to be an excellent method for including thymol in β-CD, with good encapsulation efficiency. Compared with the control, β-CD accelerated the bioavailability thymol *in vivo* and extended its half-life. Therefore, low-preservative formulations could be designed in cosmetic products.

Kayaci and Uyar [187] produced functional nanowebs containing vanillin. They prolonged the shelf life and high-temperature stability of vanillin by complexing it with CD and used an electrospinning method to impregnate PVA nanowebs with the vanillin–CD inclusion complex (vanillin–CD). To determine the most suitable CD type for the encapsulation of vanillin, they used α-CD, β-CD, and γ-CD to prepare vanillin–CD. Finally, they successfully prepared PVA–vanillin–CD nanofibers, with diameters about 200 nm, with electrospinning. The PVA–vanillin–CD nanowebs increased the durability and high-temperature stability of vanillin. The interaction between vanillin and γ-CD was stronger than that with α-CD or β-CD. This technology has potential applicability in the cosmetics industry.

Mallardo *et al.* [190] first prepared inclusion complexes of D-limonene inside β-CD cages, and then prepared poly(butylene succinate) (PBS) and inclusion complex composites by melt extrusion, using optical microscopy and infrared spectroscopy to confirm the homogeneous dispersion of the inclusion complexes in PBS. They also used TGA to show that D-limonene was thermally stabilized upon inclusion within the β-CD cages. Isothermal and nonisothermal DSC analyses showed that the inclusion complex delayed the melt crystallization of PBS. This technology has potential applicability in the preparation of the raw materials for cosmetic facial masks.

Clove oil (CO) is an aromatic oily liquid used in the cosmetics industry for its functional properties. However, it has several disadvantages, including its pungent taste, volatility, light sensitivity, and poor water solubility. Hernández-Sánchez *et al.* [180] used complexation techniques to resolve these problems. They first successfully solubilized natural CO in aqueous solution by forming inclusion complexes with β-CDs. However, phase solubility experiments showed that natural CO also forms insoluble complexes with β-CDs. Therefore, the researchers adopted a novel approach using microwave irradiation to prepare solid natural CO–β-CD complexes. They then used three different drying methods, vacuum oven drying, freeze-drying, and spray-drying, to produce the solid complexes. Freeze-drying was the best method for drying the solid CO–β-CD complexes, and microwave irradiation was used effectively to prepare the natural CO–β-CD complexes with a large yield on an industrial scale.

Abbaszadegan *et al.* [191] prepared inclusion complexes of safranal with β-CD using conventional supercritical carbon dioxide (SC-CO$_2$) and kneading (KN), coevaporation, and sealed-heating methods. They characterized the complexes with FTIR, X-ray powder diffraction, SEM, and ^{1}H-NMR spectroscopy. They also investigated the effect of the preparation method on the yield of the inclusion complex. SC-CO$_2$ was the most efficient method, and no organic solvent was involved during processing. Abbaszadegan *et al.* also investigated the effects of temperature and pressure on the properties of the complexes made with SC-CO$_2$. The formation of safranal–β-CD inclusion complexes produced with this method was confirmed with FTIR and ^{1}H-NMR. Safranal–β-CD enhanced the initial aqueous solubility of safranal by about 35%, and the inclusion complexes had a faster dissolution rate than pure safranal. This method could be used to produce the raw materials for facial masks in the cosmetics industry.

Ceborska *et al.* [192] used X-ray crystallography to examine the inclusion of (+)- and (−)-menthols in β-CD. The [2:2] complexes obtained had the typical head-to-head dimer form of β-CD. The menthol molecule in one of the CD molecules formed the dimer without docking direction and occupied two major sites in both the (+)- and (−)-menthols. The unusual "head-to-tail" array of the guest molecules was detected in both the diastereoisomeric complexes, so the two guest molecules were differently oriented inside the β-CD lumen. The researchers measured the stability constants for both complexes in solution. The solid-state complexes showed screw-channel-type packing. This study may help cosmetic technicians control the inclusion efficiency when the guest molecules have different orientations.

6.7 Conclusion

Cyclodextrins have encapsulating characteristics and can be easily derivatized, which allows the formation of inclusion complexes between CDs and various guest molecules. Currently, these CD-based complexes are mainly used for the encapsulation of specific molecules in cosmetic matrices and in the creation of new cosmetic products.

There are many applications of CDs in cosmetic formulations. Their inclusion capacity can be used to stabilize bioactive compounds; some of them use their hydrophobic cavities to load components with drawbacks, including high volatility, poor stability, poor water solubility, irritating effects, bad taste, or bad odor. These inclusion complexes can be used in almost any kind of cosmetic formulation and can play several roles simultaneously, such as in solubilization, stabilization, masking an unpleasant taste or odor, or facilitating the slow release of the guest molecule. Because they are safe for skin and humans, they are promising additives for improving the quality of cosmetic products.

Several new techniques have been developed to prepare nanofibers or films, including electrospinning and electrospraying. These electro-hydrodynamic processes are simple, cost effective, and flexible. The electrospun fibers and electrosprayed particles have many structural and functional advantages, but their applicability in the field of cosmetics is yet to be widely explored.

With the development of microneedle cosmetology, a hyaluronic acid delivery system is urgently needed. CD–skin interactions must be investigated because CDs can be used as penetration enhancers. Conventional penetration enhancers, such as alcohols and fatty acids, affect the skin barrier by damaging it, whereas hydrophilic CDs could positively affect the penetration of hyaluronic acid by increasing its availability on the skin surface. At present, various unsaturated fatty acids, fatty alcohols, and glycerol monoethers are used as penetration enhancers, but they are unstable and irritate the skin. CDs could greatly reduce these adverse effects by reducing the cutaneous penetration and therefore increasing the bioavailability of the guest molecules. CDs could provide a more efficient topical treatment by increasing the percutaneous absorption and temporarily damaging the stratum corneum barrier of the skin.

Cyclodextrins have been used in the cosmetics industry for almost 20 years. Hungarian, Japanese, and French studies were first per-formed during the mid-1990s. Since then, many products that utilize CDs for various purposes have become available on the market. In this chapter, we have mainly reviewed the CDs complexed with several different compounds with various functions, such as sunscreening, skin whitening, antiaging, moisturizing, and fragrance releasing or masking. It is useful to separate these CD functions according to the functions of their guest molecules. This can facilitate the ready acquisition of useful information from a large corpus of literature.

The design and use of CDs and their derivatives, the wide range of current research, and their potential applications will undoubtedly motivate future research.

References

1. Maugh TH. (1980). Ozone depletion would have dire effects. *Science*, 207(4429), pp. 394–395.
2. Hughes S, Bartholomew B, Hardy JC, Kramer JM. (1996). Estimates of ozone depletion and skin cancer incidence to examine the Vienna Convention achievements. *Nature*, 384(6606), pp. 256–258.
3. Jain SK, Jain NK. (2010). Multiparticulate carriers for sun-screening agents. *International Journal of Cosmetic Science*, 32(2), pp. 89–98.
4. Citernesi U. (2001). Photostability of sun filters complexed in phospholipids or β-cyclodextrin. *Cosmetics and Toiletries*, 116(9), pp. 77–86.
5. Diffey BL. (2009). Sunscreens as a preventative measure in melanoma: an evidence-based approach or the precautionary principle? *British Journal of Dermatology*, 161Suppl 3(s3), pp. 25–27.
6. Klammer H, Schlecht C, Wuttke W, Schmutzler C, Gotthardt I, Köhrle J, Jarry H. (2007). Effects of a 5-day treatment with the UV-filter octyl-methoxycinnamate (OMC) on the function of the hypothalamo-pituitary–thyroid function in rats. *Toxicology*, 238(2–3), pp. 192–199.
7. Loftsson T, Brewster ME. (1996). Pharmaceutical applications of cyclodextrins. 1. Drug solubilization and stabilization. *Journal of Pharmaceutical Sciences*, 85(10), pp. 1017–1025.
8. Rajewski RA, Stella VJ. (1996). Pharmaceutical applications of cyclodextrins. 2. *In vivo* drug delivery. *Journal of Pharmaceutical Sciences*, 85(11), pp. 1142–1169.
9. Duchêne D, Wouessidjewe D, Poelman MC. (1998). Cyclodextrins in cosmetics. *Novel Cosmetic Delivery Systems*, pp. 275–278.
10. Brewster ME, Loftsson T. (2007). Cyclodextrins as pharmaceutical solubilizers. *Advanced Drug Delivery Reviews*, 59(7), pp. 645–666.
11. Laza-Knoerr AL, Gref R, Couvreur P. (2010). Cyclodextrins for drug delivery. *Journal of Drug Targeting*, 18(9), pp. 645–656.
12. Buschmann HJ, Schollmeyer E. (2002). Applications of cyclodextrins in cosmetic products: a review. *Journal of Cosmetic Science*, 53(3), pp. 185–191.
13. Dornelas CB, Soares KCC, Castro HC, Dias LRS, Cabral LM, Vergnanini AL, Rodrigues CR. (2008). Preparation and evaluation of inclusion complexes of commercial sunscreens in cyclodextrins and montmorillonites: Performance and substantivity studies. *Drug Development and Industrial Pharmacy*, 34(5), pp. 536–546.

14. Nowakowska M, Grebosz M, Smoluch M, Tatara W. (2002). Inhibiting effect of β-cyclodextrin on thymine dimerization photosensitized by para-aminobenzoic acid. *Photochemistry and Photobiology*, 75(2), pp. 92–96.
15. Centini M, Maggiore M, Casolaro M, Andreassi M, Facino RM, Anselmi C. (2007). Cyclodextrins as cosmetic delivery systems. *Journal of Inclusion Phenomena*, 57(1), pp. 109–112.
16. Simeoni S, Scalia S, Tursilli R, Benson H. (2006). Influence of cyclodextrin complexation on the *in vitro* human skin penetration and retention of the sunscreen agent, oxybenzone. *Journal of Inclusion Phenomena*, 54(54), pp. 275–282.
17. Felton LA, Wiley CJ, Godwin DA. (2002). Influence of hydroxypropyl-β-cyclodextrin on the transdermal permeation and skin accumulation of oxybenzone. *Drug Development and Industrial Pharmacy*, 28(9), pp. 1117–1124.
18. Scalia S, Molinari A, Casolari A, Maldotti A. (2004). Complexation of the sunscreen agent, phenylbenzimidazole sulphonic acid with cyclodextrins: Effect on stability and photo-induced free radical formation. *European Journal of Pharmaceutical Sciences*, 22(4), pp. 241–249.
19. Santo S, Silvia S, Andrea B, Silvana S. (2002). Influence of hydroxypropyl-beta-cyclodextrin on photo-induced free radical production by the sunscreen agent, butyl-methoxydibenzoylmethane. *Journal of Pharmacy and Pharmacology*, 54(11), pp. 1553–1558.
20. Simeoni S, Scalia S, Benson HE. (2004). Influence of cyclodextrins on *in vitro* human skin absorption of the sunscreen, butyl-methoxydibenzoylmethane. *International Journal of Pharmaceutics*, 280(280), pp. 163–171.
21. Yang J, Wiley CJ, Godwin DA, Felton LA. (2008). Influence of hydroxypropyl-β-cyclodextrin on transdermal penetration and photostability of avobenzone. *European Journal of Pharmaceutics and Biopharmaceutics*, 69(2), pp. 605–612.
22. Inbaraj JJ, Bilski P, Chignell CF. (2002). Photophysical and photochemical studies of 2-phenylbenzimidazole and UVB sunscreen 2-phenylbenzimidazole-5-sulfonic acid. *Photochemistry and Photobiology*, 75(2), pp. 107–116.
23. Council Directive76/768/EEC of 27 July 1976 on the approximation of the laws of the Member States relating to cosmetic products, Annex VII [1976] OJ L66, p. 26.
24. Food and Drug Administration US of 21 May 1999 Sunscreen drug products for over-the-counter human use: Final monograph, Federal Register [1999]OJ L64(98), pp. 27666–27693.
25. Berset G, Gonzenbach H, Christ R, Martin R, Deflandre A, Mascotto RE, Jolley JDR, Lowell W, Pelzer R, Stiehm T. (1996). Proposed protocol for determination of photostability Part I: cosmetic UV filters. *International Journal of Cosmetic Science*, 18(4), pp. 167–177.
26. Johncock W. (1999). Sunscreen interactions in formulations. *Cosmetics and Toiletries*, 114(9), pp. 75–82.

27. Schauder S, Ippen H. (1997). Contact and photocontact sensitivity to sunscreens. *Contact Dermatitis*, 37(5), pp. 221–232.
28. Loftsson T. (1998). Increasing the cyclodextrin complexation of drugs and drug biovailability through addition of water-soluble polymers. *Pharmazie*, 53(11), pp. 733–740.
29. Scalia S, Villani S, Casolari A. (1999). Inclusion complexation of the sunscreen agent 2-ethylhexyl- p -dimethylaminobenzoate with hydroxypropyl-β-cyclodextrin: effect on photostability. *Journal of Pharmacy and Pharmacology*, 51(51), pp. 1367–1374.
30. Loftsson T, Masson M. (2001). Cyclodextrins in topical drug formulations: theory and practice. *International Journal of Pharmaceutics*, 225(1–2), pp. 15–30.
31. Treffel P, Gabard B. (1996). Skin penetration and sun protection factor of ultraviolet filters from two vehicles. *Pharmaceutical Research*, 13(5), pp. 770–774.
32. Stevenson C, Davies R. (1999). Photosensitization of guanine-specific DNA damage by 2-phenylbenzimidazole and the sunscreen agent 2-phenylbenzimidazole-5-sulfonic acid. *Chemical Research in Toxicology*, 12(1), pp. 38–45.
33. Buettner GR. (1987). Spin trapping: ESR parameters of spin adducts. *Free Radical Biology & Medicine*, 3(4), pp. 259–303.
34. Beauchamp C. (1970). A mechanism for the production of ethylene from methional: the generation of the hydroxyl radical by xanthine oxidase. *Journal of Biological Chemistry*, 245(18), pp. 4641–4646.
35. Ross L, Barclay C, Artz JD, Mowat JJ. (1995). Partitioning and antioxidant action of the water-soluble antioxidant, Trolox, between the aqueous and lipid phases of phosphatidylcholine membranes: 14 C tracer and product studies. *Biochimica and Biophysica Acta*, 1237(1), pp. 77–85.
36. Carlotti ME, Sapino S, Vione D, Pelizzetti E, Trotta M. (2004). Photostability of Trolox in water/ethanol, water, and Oramix CG 110 in the absence and in the presence of TiO$_2$. *Journal of Dispersion Science and Technology*, 25(2), pp. 193–207.
37. Nonell S, Moncayo L, Trull F, Amat-Guerri F, Lissi EA, Soltermann AT, Criado S, García NA. (1995). Solvent influence on the kinetics of the photodynamic degradation of trolox, a water-soluble model compound for vitamin E. *Journal of Photochemistry and Photobiology B: Biology*, 29(2), pp. 157–162.
38. Higuchi TA, Connors KA. (1965). Phase-solubility techniques. *Advances in Analytical Chemistry and Instrumentation*, 54, pp. 212–217.
39. Másson M, Loftsson T, Másson GS, Stefánsson E. (1999). Cyclodextrins as permeation enhancers: some theoretical evaluations and *in vitro* testing. *Journal of Controlled Release*, 59(1), pp. 107–118.
40. Carlotti ME, Sapino S, Marino S, Ugazio E, Trotta F, Vione D, Chirio D, Cavalli R. (2008). Influence of hydroxypropyl-β-cyclodextrin on the photostability and antiradical activity of Trolox. *Journal of Inclusion Phenomena and Macrocyclic Chemistry*, 61(3–4), pp. 279–287.
41. Canipelle M, Tilloy S, Ponchel A, Bricout H, Monflier E. (2005). Complexation of monosulfonated triphenylphosphine with chemically modified

β-cyclodextrins: effect of substituents on the stability of inclusion complexes. *Journal of Inclusion Phenomena*, 51(1–2), pp. 79–85.

42. Sapino S, Trotta M, Ermondi G, Caron G, Cavalli R, Carlotti ME. (2008). On the complexation of Trolox with methyl-β-cyclodextrin: characterization, molecular modelling and photostabilizing properties. *Journal of Inclusion Phenomena and Macrocyclic Chemistry*, 62(1–2), pp. 179–186.

43. Coppini D, Paganizzi P, Santi P, Ghirardini A. (2001). Capacità protettiva nei confronti delle radiazioni solari di derivati di origine vegetale. *Cosmetic News*, 136, pp. 15–20.

44. Xu Z, Godber JS. (1999). Purification and identification of components of γ-oryzanol in rice bran oil. *Journal of Agricultural and Food Chemistry*, 47(7), pp. 2724–2728.

45. Cicero AFG, Gaddi A. (2001). Rice bran oil and γ-oryzanol in the treatment of hyperlipoproteinaemias and other conditions. *Phytotherapy Research*, 15(4), pp. 277–289.

46. Cavalli R, Trotta F, Tumiatti W. (2006). Cyclodextrin-based nanosponges for drug delivery. *Journal of Inclusion Phenomena*, 56(1), pp. 209–213.

47. Carlotti ME, Sapino S, Vione D, Minero C, Peira E, Trotta M. (2007). Study on the photodegradation of salicylic acid in different vehicles in the absence and in the presence of TiO₂. *Journal of Dispersion Science and Technology*, 28(5), pp. 805–818.

48. Godwin DA, Wiley CJ, Felton LA. (2006). Using cyclodextrin complexation to enhance secondary photoprotection of topically applied ibuprofen. *European Journal of Pharmaceutics and Biopharmaceutics*, 62(1), pp. 85–93.

49. Sapino S, Carlotti ME, Caron G, Ugazio E, Cavalli R. (2009). In silico design, photostability and biological properties of the complex resveratrol/ hydroxypropyl-β-cyclodextrin. *Journal of Inclusion Phenomena*, 63(1), pp. 171–180.

50. Monti D, Tampucci S, Chetoni P, Burgalassi S, Saino V, Centini M, Staltari L, Anselmi C. (2011). Permeation and distribution of ferulic acid and its α-cyclodextrin complex from different formulations in hairless rat skin. *AAPS PharmSciTech*, 12(2), pp. 514–520.

51. MHW Ministry of Health and Welfare. (2000). Notification No. 331/2000, Standards for Cosmetics, Retrieved from: http://www.mhlw.go.jp/english/topics/cosmetics/index.html.

52. Anselmi C, Centini M, Ricci M, Buonocore A, Granata P, Tsuno T, Facino RM. (2006). Analytical characterization of a ferulic acid/γ-cyclodextrin inclusion complex. *Journal of Pharmaceutical and Biomedical Analysis*, 40(4), pp. 875–881.

53. Scalia S, Tursilli R, Iannuccelli V. (2007). Complexation of the sunscreen agent, 4-methylbenzylidene camphor with cyclodextrins: effect on photostability and human stratum corneum penetration. *Journal of Pharmaceutical and Biomedical Analysis*, 44(1), pp. 29–34.

54. Lowe NJ. (1996). *Sunscreens: Development, Evaluation, and Regulatory Aspects*. Boca Raton: CRC Press.

55. Anselmi C, Centini M, Maggiore M, Gaggelli N, Andreassi M, Buonocore A, Beretta G, Facino RM. (2008). Non-covalent inclusion of ferulic acid with α-cyclodextrin improves photo-stability and delivery: NMR and modeling studies. *Journal of Pharmaceutical and Biomedical Analysis*, 46(4), pp. 645–652.

56. Casolaro M, Anselmi C, Picciocchi G. (2005). The protonation thermodynamics of ferulic acid/γ-cyclodextrin inclusion compounds. *Thermochimica Acta*, 425(1), pp. 143–147.

57. Challa R, Ahuja A, Ali J, Khar R. (2005). Cyclodextrins in drug delivery: an updated review. *AAPS PharmSciTech*, 6(2), pp. E329–E357.

58. Pekkarinen SS, Stöckmann H, Schwarz K, Heinonen IM, Hopia AI. (1999). Antioxidant activity and partitioning of phenolic acids in bulk and emulsified methyl linoleate. *Journal of Agricultural and Food Chemistry*, 47(8), pp. 3036–3043.

59. Arbab A, Eltahir M. (2010). Review on skin whitening agents. *Khartoum Pharmacy Journal*, 13(1), pp. 5–9.

60. Huixia Q, Xiaohui L, Chengda Y, Yanlu Z, Senee J, Laurent A, Bazin R, Flament F, Adam A, Piot B. (2011). Instrumental and clinical studies of the facial skin tone and pigmentation of Shanghainese women. Changes induced by age and a cosmetic whitening product. *International Journal of Cosmetic Science*, 34(1), pp. 49–54.

61. Smit N, Vicanova J, Pavel S. (2009). The hunt for natural skin whitening agents. *International Journal of Molecular Sciences*, 10(12), pp. 5326–5349.

62. Hearing VJ, Tsukamoto K. (1991). Enzymatic control of pigmentation in mammals. *The FASEB Journal*, 5(14), pp. 2902–2909.

63. Siddique S, Parveen Z, Ali Z, Zaheer M. (2012). Qualitative and quantitative estimation of hydroquinone in skin whitening cosmetics. *Journal of Cosmetics Dermatological Sciences and Applications*, 2(3), p. 224.

64. Boonme P, Junyaprasert VB, Suksawad N, Songkro S. (2009). Microemulsions and nanoemulsions: Novel vehicles for whitening cosmeceuticals. *Journal of Biomedical Nanotechnology*, 5(4), pp. 373–383.

65. Chang TS. (2009). An updated review of tyrosinase inhibitors. *International Journal of Molecular Sciences*, 10(6), pp. 2440–2475.

66. Solano F, Briganti S, Picardo M, Ghanem G. (2006). Hypopigmenting agents: an updated review on biological, chemical and clinical aspects. *Pigment Cell Research*, 19(6), pp. 550–571.

67. Wiechers JW, Rawlings AV, Garcia C, Chesné C, Balaguer P, Nicolas JC, Corre S, Galibert MD. (2005). A new mechanism of action for skin whitening agents: binding to the peroxisome proliferator-activated receptor 1. *International Journal of Cosmetic Science*, 27(2), pp. 123–132.

68. Barel AO, Paye M, Maibach HI. (2015). *Handbook of Cosmetic Science and Technology*, Fourth Edition. Boca Raton: CRC Press.

69. Gupta SK, Gautam A, Kumar S. (2014). Natural skin whitening agents: acurrent status. *Advances in Biological Research*, 8(6), pp. 257–259.

70. Ullah S, Son S, Yun HY, Kim DH, Chun P, Moon HR. (2016). Tyrosinase inhibitors: a patent review (2011–2015). *Expert Opinion on Therapeutic Patents*, 26(3), pp. 347–362.
71. Shigeta Y, Imanaka H, Ando H, Ryu A, Oku N, Baba N, Makino T. (2004). Skin whitening effect of linoleic acid is enhanced by liposomal formulations. *Biological and Pharmaceutical Bulletin*, 27(4), pp. 591–594.
72. Khan NA, Durakshan M. (2013). Cyclodextrin: Anoverview. *International Journal of Bioassays*, 2(6), pp. 858–865.
73. Lucas-Abellán C, Fortea I, López-Nicolás JM, Núñez-Delicado E. (2007). Cyclodextrins as resveratrol carrier system. *Food Chemistry*, 104(1), pp. 39–44.
74. Hădărugă NG, Hădărugă DI, Păunescu V, Tatu C, Ordodi VL, Bandur G, Lupea AX. (2006). Thermal stability of the linoleic acid/α- and β-cyclodextrin complexes. *Food Chemistry*, 99(3), pp. 500–508.
75. López-Nicolás JM, Bru R, Sánchez-Ferrer A, García-Carmona F. (1995). Use of 'soluble lipids' for biochemical processes: linoleic acid-cyclodextrin inclusion complexes in aqueous solutions. *Biochemical Journal*, 308(1), pp. 151–154.
76. Bochot A, Piel G. (2011). Applications of cyclodextrins for skin formulation and delivery. In: Bilensoy E (Ed.), *Cyclodextrins in Pharmaceutics, Cosmetics, and Biomedicine: Current and Future Industrial Applications*. Hoboken: John Wiley & Sons, pp. 159–176.
77. Hougeir FG, Kircik L. (2012). A review of delivery systems in cosmetics. *Dermatologic Therapy*, 25(3), pp. 234–237.
78. Duchêne D, Bochot A, Yu SC, Pépin C, Seiller M. (2003). Cyclodextrins and emulsions. *International Journal of Pharmaceutics*, 266(1–2), pp. 85–90.
79. Cal K, Centkowska K. (2008). Use of cyclodextrins in topical formulations: practical aspects. *European Journal of Pharmaceutics and Biopharmaceutics*, 68(3), pp. 467–478.
80. Pinho E, Grootveld M, Soares G, Henriques M. (2014). Cyclodextrins as encapsulation agents for plant bioactive compounds. *Carbohydrate Polymers*, 101(1), pp. 121–135.
81. Tarimci N. (2011). Cyclodextrins in the cosmetic field. In: Bilensoy E (Ed.), *Cyclodextrins in Pharmaceuticals, Cosmetics, and Biomedicine: Current and Future Industrial Applications*. Hoboken: John Wiley & Sons, pp. 131–144.
82. Valle EMMD. (2004). Cyclodextrins and their uses: a review. *Process Biochemistry*, 39(9), pp. 1033–1046.
83. Patravale V, Mandawgade S. (2008). Novel cosmetic delivery systems: an application update. *International Journal of Cosmetic Science*, 30(1), pp. 19–33.
84. Centini M, Maggiore M, Casolaro M, Andreassi M, Facino RM, Anselmi C. (2007). Cyclodextrins as cosmetic delivery systems. *Journal of Inclusion Phenomena*, 57(1), pp. 109–112.
85. Gillbro JM, Olsson MJ. (2011). The melanogenesis and mechanisms of skin-lightening agents — existing and new approaches. *International Journal of Cosmetic Science*, 33(3), pp. 210–221.

86. Zhong L, Bo C, Hu Y, Zhang Y, Zou G. (2009). Complexation of resveratrol with cyclodextrins: solubility and antioxidant activity. *Food Chemistry*, 113(1), pp. 17–20.

87. Petit L, Piérard GE. (2003). Skin-lightening products revisited. *International Journal of Cosmetic Science*, 25(4), pp. 169–181.

88. Hatae S, Nakashima K. (1989). Whitening cosmetic. US Patent 4,847,074,11 Jul 1989.

89. Núñezdelicado E, Alvaro Sánchezferrer A, Garcíacarmona F. (1997). Cyclodextrins as secondary antioxidants:? synergism with ascorbic acid. *Journal of Agricultural and Food Chemistry*, 45(8), pp. 2830–2835.

90. Davaran S, Hanaee J, Rashidi MR, Valiolah F, Hashemi M. (2006). Influence of poly(ethylene glycol)-α-cyclodextrin complexes on stabilization and transdermal permeation of ascorbic acid. *Journal of Biomedical Materials Research Part A*, 78A(3), pp. 590–594.

91. Lin JW, Chiang HM, Lin YCWen KC. (2008). Natural products with skin — whitening effects. *Journal of Food and Drug Analysis*, 16(2), pp. 1–10.

92. Parvez S, Kang M, Chung HS, Cho C, Hong MC, Shin MK, Bae H. (2006). Survey and mechanism of skin depigmenting and lightening agents. *Phytotherapy Research*, 20(11), pp. 921–934.

93. Lucasabellán C, Fortea I, Gabaldón JA, Núñezdelicado E. (2008). Encapsulation of quercetin and myricetin in cyclodextrins at acidic pH. *Journal of Agricultural and Food Chemistry*, 56(1), pp. 255–259.

94. Mercader-Ros MT, Lucas-Abellán C, Fortea MI, Gabaldón JA, Núñez-Delicado E. (2010). Effect of HP-β-cyclodextrins complexation on the antioxidant activity of flavonols. *Food Chemistry*, 118(3), pp. 769–773.

95. Vyas A, Saraf S, Saraf S. (2008). Cyclodextrin based novel drug delivery systems. *Journal of Inclusion Phenomena*, 62(1), pp. 23–42.

96. Lastra CADL, Villegas I. (2005). Resveratrol as an anti-inflammatory and anti-aging agent: mechanisms and clinical implications. *Molecular Nutrition and Food Research*, 49(5), pp. 405–430.

97. Cheng SY, Yuen CWM, Kan CW, Cheuk KKL. (2008). Development of cosmetic textiles using microencapsulation technology. *Research Journal of Textile and Apparel*, 12(4), pp. 41–51.

98. Shimoda K, Akagi M, Hamada H. (2009). Production of β-maltooligosaccharides of α-and δ-tocopherols by *Klebsiella pneumoniae* and cyclodextrin glucanotransferase as anti-allergic agents. *Molecules*, 14(8), pp. 3106–3114.

99. Benita S. (2005). *Microencapsulation: Methods and Industrial Applications*, Second Edition. Boca Raton: CRC Press.

100. Gonzalez S, Gilaberte Y, Philips N, Juarranz A. (2010). Fernblock, a nutriceutical with photoprotective properties and potential preventive agent for skin photoaging and photoinduced skin cancers. *International Journal of Molecular Sciences*, 12(12), pp. 8466–8475.

101. Puppo MC, Speroni F, Chapleau N, Lamballerie MD, Añón MC, Anton M. (2005). Effect of high-pressure treatment on emulsifying properties of soybean proteins. *Food Hydrocolloids*, 19(2), pp. 289–296.

102. Li H, Xu X, Liu M, Sun D, Li L. (2010). Microcalorimetric and spectrographic studies on host–guest interactions of α-, β-, γ- and mβ-cyclodextrin with resveratrol. *Thermochimica Acta*, 510(1), pp. 168–172.

103. Gellman S, Rozema DB. (1997). Method for refolding misfolded enzymes with detergent and cyclodextrin. *Biotechnology Advances*, 15(2), pp. 515–515(1).

104. Reynaldo V, Cao R, Alex F. (2007). Supramolecular chemistry of cyclodextrins in enzyme technology. *Chemical Reviews*, 107(7), pp. 3088–3116.

105. Frémont L. (2000). Biological effects of resveratrol. *Life Sciences*, 66(8), pp. 663–673.

106. Bhat KPL, Pezzuto JM. (1999). Cancer chemopreventive activity of resveratrol. *Annals of the New York Academy of Sciences*, 957(1), pp. 210–229.

107. Pace-Asciak CR, Hahn S, Diamandis EP, Soleas G, Goldberg DM. (1995). The red wine phenolics trans-resveratrol and quercetin block human platelet aggregation and eicosanoid synthesis: implications for protection against coronary heart disease. *Clinica Chimica Acta*, 235(2), pp. 207–219.

108. Calabrò ML, Tommasini S, Donato P, Raneri D, Stancanelli R, Ficarra P, Ficarra R, Costa C, Catania S, Rustichelli C. (2004). Effects of α- and β-cyclodextrin complexation on the physico-chemical properties and antioxidant activity of some 3-hydroxyflavones. *Journal of Pharmaceutical and Biomedical Analysis*, 35(2), pp. 365–377.

109. Karathanos VT, Mourtzinos I, Yannakopoulou K, Andrikopoulos NK. (2007). Study of the solubility, antioxidant activity and structure of inclusion complex of vanillin with β-cyclodextrin. *Food Chemistry*, 101(2), pp. 652–658.

110. López-Nicolás JM, Núñez-Delicado E, Pérez-López AJ, Ángel Carbonell Bar-rachina, Cuadra-Crespo P. (2006). Determination of stoichiometric coefficients and apparent formation constants for β-cyclodextrin complexes of trans-resveratrol using reversed-phase liquid chromatography. *Journal of Chromatography A*, 1135(2), pp. 158–165.

111. Jeandet P, Douillet-Breuil AC, Bessis R, Debord S, Sbaghi M, Adrian M. (2002). Phytoalexins from the vitaceae: biosynthesis, phytoalexin gene expression in transgenic plants, antifungal activity, and metabolism. *Journal of Agricultural and Food Chemistry*, 50(10), pp. 2731–2741.

112. Siemann EH, Creasy LL. (1992). Concentration of the phytoalexin resveratrol in wine. *American Journal of Enology and Viticulture*, 43(1), pp. 49–52.

113. Bradamante S, Barenghi L, Villa A. (2004). Cardiovascular protective effects of resveratrol. *Cardiovascular Drug Reviews*, 22(3), pp. 169–188.

114. Almagro L, Sabater-Jara AB, Belchí-Navarro S, Fernández-Pérez F, Bru R, Pedreño MA. (2011). Effect of UV light on secondary metabolite biosynthesis in plant cell cultures elicited with cyclodextrins and methyljasmonate. In: Vasanthaiah HKN, Kambiranda D (Eds.), *Plants and Environment*. Rijeka: Intech, pp. 115–136.

115. Lijavetzky D, Almagro L, Belchi-Navarro S, Martínez-Zapater JM, Bru R, Pedreño MA. (2008). Synergistic effect of methyljasmonate and cyclodextrin on stilbene biosynthesis pathway gene expression and resveratrol production in Monastrell grapevine cell cultures. *BMC Research Notes*, 1(1), p. 132.

116. Regiert M. (2007). Oxidation-stable linoleic acid by inclusion in α-cyclo-dextrin. *Journal of Inclusion Phenomena*, 57(1), pp. 471–474.
117. Lautenschlager I, Höckerstedt K, Taskinen E. (2003). Histologic findings associated with cmv infection in liver transplantation. *Transplantation Proceedings*, 35(2), p. 819.
118. Astray G, Gonzalez-Barreiro C, Mejuto JC, Rial-Otero R, Simal-Gándara J. (2009). A review on the use of cyclodextrins in foods. *Food Hydrocolloids*, 23(7), pp. 1631–1640.
119. Szejtli J. (1997). Utilization of cyclodextrins in industrial products and processes. *Journal of Materials Chemistry*, 7(4), pp. 575–587.
120. Timsina B, Shukla M, Nadumane VK. (2012). A review of few essential oils and their anticancer property. *Journal of Natural Pharmaceuticals*, 3(1), pp. 1–8.
121. Tsai ML, Lin CC, Lin WC, Yang CH. (2011). Antimicrobial, antioxidant, and anti-inflammatory activities of essential oils from five selected herbs. *Bioscience Biotechnology and Biochemistry*, 75(10), pp. 1977–1983.
122. Marques HMC. (2010). A review on cyclodextrin encapsulation of essential oils and volatiles. *Flavour and Fragrance Journal*, 25(5), pp. 313–326.
123. Kopecký F, Kopecká B, Kaclík P. (2001). Solubility study of nimodipine inclusion complexation with α- and β-cyclodextrin and some substituted cyclodextrins. *Journal of Inclusion Phenomena*, 39(3), pp. 215–217.
124. Numanoğlu U, Şen T, Tarimci N, Kartal M, Koo OMY, Önyüksel H. (2006). Use of cyclodextrins as a cosmetic delivery system for fragrance materials: linalool and benzyl acetate. *AAPS PharmSciTech*, 8(4), pp. 34–42.
125. Yuan C, Jin Z, Li X. (2008). Evaluation of complex forming ability of hydroxypropyl-β-cyclodextrins. *Food Chemistry*, 106(1), pp. 50–55.
126. Saito Y, Tanemura I, Sato T, Ueda H. (1999). Interaction of fragrance materials With 2-hydroxypropyl-β-cyclodextrin by static and dynamic headspace methods. *International Journal of Cosmetic Science*, 21(3), pp. 189–198.
127. Aicart E, Junquera E. (2003). Complex formation between purine derivatives and cyclodextrins: a fluorescence spectroscopy study. *Journal of Inclusion Phenomena*, 47(3), pp. 161–165.
128. Dall'Asta C, Ingletto G, Corradini R, Galaverna G, Marchelli R. (2003). Fluorescence enhancement of alatoxins using native and substituted cyclodextrins. *Journal of Inclusion Phenomena*, 45(3–4), pp. 257–263.
129. Jiang H, Zhang S, Cui Y, Xie Y, Shi Q. (2011). NMR and fluorescence enhancement effect studies of ternary inclusion complexes among naphthalenediamines-triethylene tetramine modified β-cyclodextrin-lanthanide metal ions. *African Journal of Pure and Applied Chemistry*, 5(6), pp. 136–144.
130. Liu Y, Yang YW, Zhao Y, Li L, Zhang HY, Kang SZ. (2003). Molecular recognition and cooperative binding ability of fluorescent dyes by bridged Bisβ-cyclodextrin)s tethered with aromatic diamine. *Journal of Inclusion Phenomena*, 47(3), pp. 155–160.
131. Easton C, Lincoln S. (1999). *Modified Cyclodextrins: Scaffolds and Templates for Supramolecular Chemistry*. London: Imperial College Press.

132. Gazpio C, Sánchez M, Zornoza A, Martıìn C, Martıìnez-Ohárriz C, Vélaz I. (2003). A fluorimetric study of pindolol and its complexes with cyclodextrins. *Talanta*, 60(2–3), pp. 477–482.
133. Kfoury M, Auezova L, Fourmentin S, Greige-Gerges H. (2014). Investigation of monoterpenes complexation with hydroxypropyl-β-cyclodextrin. *Journal of Inclusion Phenomena and Macrocyclic Chemistry*, 80(1–2), pp. 51–60.
134. Mourtzinos I, Kalogeropoulos N, Papadakis SE, Konstantinou K, Karathanos VT. (2008). Encapsulation of nutraceutical monoterpenes in β-cyclodextrin and modified starch. *Journal of Food Science*, 73(1), pp. 1679–1685.
135. Gajare P, Patil C, Kalyane N, Pore Y. (2009). Effect of hydrophilic polymers on pioglitazone complexation with hydroxypropyl-β-cyclodextrin. *Digest Journal of Nanomaterials and Biostructures*, 4(4), pp. 891–897.
136. Shukla D, Chakraborty S, Singh S, Mishra B. (2009). Preparation and *in vitro* characterization of Risperidone-cyclodextrin inclusion complexes as a potential injectable product. *DARU Journal of Pharmaceutical Sciences*, 17(4), pp. 226–235.
137. Kaur LP, Sharma S, Guleri TK. (2013). Microencapsulation: A new era in noval drug delivery. *The International Journal of Pharmaceutical Research and Bio-Science*, 2(2), pp. 456–468.
138. Martins IM, Barreiro MF, Coelho M, Rodrigues AE. (2014). Microencapsulation of essential oils with biodegradable polymeric carriers for cosmetic applications. *Chemical Engineering Journal*, 245(6), pp. 191–200.
139. Rosen MR. (2005). *Delivery System Handbook for Personal Care and Cosmetic Products: Technology, Applications and Formulations*. Norwich: William Andrew Publishing.
140. Suraweera RK, Pasansi HGP, Herath HMDR, Wickramaratne DBM, Sudeshika SHT, Niyangoda D, *et al.* (2014). Formulation and stability evaluation of ketoprofen loaded virgin coconut oil based creamy emulsion. *International Journal of Pharmacy and Pharmaceutical Sciences*, 6(8), pp. 249–254.
141. Flynn GL. (2002). Cutaneous and transdermal delivery-processes and systems of delivery. In: Banker GS, Rhodes CT (Eds.), *Modern Pharmaceutics*, Fourth Edition. Boca Raton: CRC Press, pp. 293–364.
142. Wiechers JW. (2008). *Science and Applications of Skin Delivery Systems*. Carol Stream: Allured Publishing.
143. Haddadi A, Aboofazeli R, Erfan M, Farboud ES. (2008). Topical delivery of urea encapsulated in biodegradable PLGA microparticles: O/W and W/O creams. *Journal of Microencapsulation*, 25(6), pp. 379–386.
144. Harding CR. (2004). The stratum corneum: structure and function in health and disease. *Dermatologic Therapy*, 17(s1), pp. 6–15.
145. Rein H. (1924). Experimental electroendosmotic studies on living human skin. *Zeitschrift fur Biologie*, 81, p. 124.
146. Cattaneo M. (2005). Topical delivery systems based on polysaccharide microspheres. In: Rosen MR (Ed.), *Delivery System Handbook for Personal Care and Cosmetic Products: Technology, Applications, and Formulations*. Norwich: William Andrew Publishing, pp. 273–282.

147. Fu X, Ping Q, Gao Y. (2005). Effects of formulation factors on encapsulation efficiency and release behaviour *in vitro* of huperzine A-PLGA microspheres. *Journal of Microencapsulation*, 22(7), pp. 705–714.

148. Abla M, Banga A. (2013). Quantification of skin penetration of antioxidants of varying lipophilicity. *International Journal of Cosmetic Science*, 35(1), pp. 19–26.

149. Cohen KA, Ross D, Suss H. (1999). Skin revitalizing makeup composition, US, US 5876736 A.

150. Sao Pedro A, Cabral-Albuquerque E, Ferreira D, Sarmento B. (2009). Chitosan: An option for development of essential oil delivery systems for oral cavity care? *Carbohydrate Polymers*, 76(4), pp. 501–508.

151. Wille JJ. (2006). Thixogel: a starch matrix encapsulation technology for topical drug and cosmetic delivery. In: Wille JJ (Ed.), *Skin Delivery Systems: Transdermals, Dermatologicals and Cosmetic Actives*, pp. 223–245.

152. Ammala A. (2013). Biodegradable polymers as encapsulation materials for cosmetics and personal care markets. *International Journal of Cosmetic Science*, 35(2), pp. 113–124.

153. Mishra N, Goyal AK, Khatri K, Vaidya B, Paliwal R, Rai S, Mehta A, Tiwari S, Vyas S, Vyas SP. (2008). Biodegradable polymer based particulate carrier (s) for the delivery of proteins and peptides. *Anti-Inflammatory and Anti-Allergy Agents in Medicinal Chemistry*, 7(4), pp. 240–251.

154. Stevanović M, Savić J, Jordović B, Uskoković D. (2007). Fabrication, *in vitro* degradation and the release behaviours of poly (DL-lactide-co-glycolide) nanospheres containing ascorbic acid. *Colloids and Surfaces B: Biointerfaces*, 59(2), pp. 215–223.

155. Lam P, Gambari R. (2014). Advanced progress of microencapsulation technologies: *in vivo* and *in vitro* models for studying oral and transdermal drug deliveries. *Journal of Controlled Release*, 178, pp. 25–45.

156. Nack H. (1970). Microencapsulation techniques applications and problems. *Journal of the Society of Cosmetic Chemists*, 21, pp. 85–98.

157. Handjani RM, Kauffmann M, Huguenin F. (1993). U.S. Patent No. 5,204,111. Washington, DC: U.S. Patent and Trademark Office.

158. European Commission. (2013). Growth, Internal Market, Industry, Entrepreneurship and SMEs. Retrieved from: http://ec.europa.eu/growth/sectors/cosmetics/legislation.

159. Vögtle F, Alfter F. (1993). *Supramolecular Chemistry: An Introduction*. New York: John Wiley & Sons.

160. Muñoz-Botella S, Del Castillo B, Martyn M. (1995). Cyclodextrin properties and applications of inclusion complex formation. *Ars Pharmaceutica*, 36(2), pp. 187–198.

161. Szejtli J. (1998). Introduction and general overview of cyclodextrin chemistry. *Chemical Reviews*, 98(5), pp. 1743–1754.

162. Perugini P, Genta I, Pavanetto F, Conti B, Scalia S, Baruffini A. (2000). Study on glycolic acid delivery by liposomes and microspheres. *International Journal of Pharmaceutics*, 196(1), pp. 51–61.

163. Estanqueiro M, Conceição J, Amaral MH, Lobo JMS. (2014). Solid lipid nanoparticles and nanostructured lipid carriers in moisturizing cosmetics. *Household and Personal Care Today*, 9(5), pp. 43–47.
164. Holland L, Rizzi G, Malton P. (2000). Cosmetic compositions comprising cyclic oligosaccharides and fragrance. PCT Int Appl WO, Patent No. 67, 716.
165. Zhu GY, Xiao ZB, Zhou RJ, Feng NN. (2015). Production of a transparent lavender flavour nanocapsule aqueous solution and pyrolysis characteristics of flavour nanocapsule. *Journal of Food Science and Technology*, 52(7), pp. 4607–4612.
166. Ciobanu A, Mallard I, Landy D, Brabie G, Nistor D, Fourmentin S. (2012). Inclusion interactions of cyclodextrins and crosslinked cyclodextrin polymers with linalool and camphor in *Lavandula angustifolia* essential oil. *Carbohydrate Polymers*, 87, pp. 1963–1970.
167. Numanoglu U, Sen T, Tarimci N, Kartal M, Koo OMY, Onyuksel H. (2007). Use of cyclodextrins as a cosmetic delivery system for fragrance materials: linalool and benzyl acetate. *AAPS PharmSciTech*, 8(4), pp. 34–42.
168. Li Y, Huang YQ, Fan HF, Xia Q. (2014). Comparison of release behaviors of fragrance/hydroxypropyl-β-cyclodextrin inclusion complex and fragrance microcapsules. *Integrated Ferroelectrics*, 152(1), pp. 81–89.
169. Ciobanu A, Ruellan S, Mallard I, Landy D, Gennequin C, Siffert S, Fourmentin S. (2013). Cyclodextrin-intercalated layered double hydroxides for fragrance release. *Journal of Inclusion Phenomena and Macrocyclic Chemistry*, 75, pp. 333–339.
170. Liu CJ, Hayashi K. (2014). Visualization of controlled fragrance release from cyclodextrin inclusion complexes by fluorescence imaging. *Flavour and Fragrance Journal*, 29, pp. 356–363.
171. Kfoury M, Auezova L, Fourmentin S, Greige-Gerges H. (2014). Investigation of monoterpenes complexation with hydroxypropyl-β-cyclodextrin. *Journal of Inclusion Phenomena and Macrocyclic Chemistry*, 80(1), pp. 51–60.
172. Menezes PP, Serafini MR, Quintans-Júnior LJ, *et al.* (2014). Inclusion complex of (−)-linalool and β-cyclodextrin. *Journal of Thermal Analysis and Calorimetry*, 115(3), pp. 2429–2437.
173. Yang ZJ, Xiao ZB, Ji HB. (2015). Solid inclusion complex of terpinen-4-ol/β-cyclodextrin: kinetic release, mechanism and its antibacterial activity. *Flavour and Fragrance Journal*, 30(2), pp. 179–187.
174. Lopez MD, Maudhuit A, Pascual-Villalobos MJ, Poncelet D. (2012). Development of formulations to improve the controlled-release of linalool to be applied as an insecticide. *Journal of Agricultural and Food Chemistry*, 60(5), pp. 1187–1192.
175. Karathanos VT, Mourtzinos I, Yannakopoulou K, Andrikopoulos NK. (2007). Study of the solubility, antioxidant activity and structure of inclusion complex of vanillin with β-cyclodextrin. *Food Chemistry*, 101(2), pp. 652–658.
176. Ji HB, Long QP, Chen HY, Zhou XT, Hu XF. (2011). β-Cyclodextrin inclusive interaction driven separation of organic compounds. *AIChE Journal*, 57(9), pp. 2341–2352.

177. Sebaaly C, Charcosset C, Stainmesse S, Fessi H, Greige-Gerges H. (2016). Clove essential oil-in-cyclodextrin-in-liposomes in the aqueous and lyophilized states: From laboratory to large scale using a membrane contactor. *Carbohydrate Polymers*, 138, pp. 75–85.

178. Abarca RL, Rodriguez FJ, Guarda A, Galotto MJ, Bruna JE. (2016). Characterization of beta-cyclodextrin inclusion complexes containing an essential oil component. *Food Chemistry*, 196(1), pp. 968–975.

179. Hernández-Sánchez P, López-Miranda S, Guardiola L, Serrano-Martínez A, Antonio Gabaldón J, Nuñez-Delicado E. (2016). Optimization of a method for preparing solid complexes of essential clove oil with β-cyclodextrins. *Journal of the Science of Food and Agriculture*, in press. doi: 10.1002/jsfa.7781.

180. Aytac Z, Yildiz ZI, Kayaci-Senirmak F, Keskin NOS, Tekinay T, Uyar T. (2016). Electrospinning of polymer-free cyclodextrin/geraniol–inclusion complex nanofibers: Enhanced shelf-life of geraniol with antibacterial and antioxidant properties. *RSC Advances*, 6, pp. 46089–46099.

181. Higueras L, Lopez-Carballo G, Gavara R, Hernendez-Muaoz P. (2015). Incorporation of hydroxypropyl-β-cyclodextrins into chitosan films to tailor loading capacity for active aroma compound carvacrol. *Food Hydrocolloids*, 43, pp. 603–611.

182. Mascheroni E, Fuenmayor CA, Cosio MS, Di Silvestro G, Piergiovanni L, Mannino S, Schiraldi A. (2013). Encapsulation of volatiles in nanofibrous polysaccharide membranes for humidity-triggered release. *Carbohydrate Polymers*, 98(1), pp. 17–25.

183. Baránková E, Dohnal V. (2016). Effect of additives on volatility of aroma compounds from dilute aqueous solutions. *Fluid Phase Equilibria*, 407, pp. 217–223.

184. Ciobanu A, Landy D, Fourmentin S. (2013). Complexation efficiency of cyclodextrins for volatile flavor compounds. *Food Research International*, 53, pp. 110–114.

185. Kfoury M, Auezova L, Ruellan S, Greige-Gerges H, Fourmentin S. (2015). Complexation of estragole as pure compound and as main component of basil and tarragon essential oils with cyclodextrins. *Carbohydrate Polymers*, 118, pp. 156–164.

186. Kayaci F, Ertas Y, Uyar T. (2013). Enhanced thermal stability of eugenol by cyclodextrin inclusion complex encapsulated in electrospun polymeric nanofibers. *Journal of Agricultural and Food Chemistry*, 61(34), pp. 8156–8165.

187. Xiao Z, Feng N, Zhu G, Niu Y. (2015). Preparation and application of citral-monochlorotriazine-β-cyclodextrin inclusion complex nanocapsule. *The Journal of The Textile Institute*, 107(1), pp. 64–71.

188. Kfoury M, Auezova L, Greige-Gerges H, Ruellan S, Fourmentin S. (2014). Cyclodextrin, an efficient tool for trans-anethole encapsulation: Chromatographic, spectroscopic, thermal and structural studies. *Food Chemistry*, 164, pp. 454–461.

189. Nieddu M, Rassu G, Boatto G, Bosi P, Trevisi P, Giunchedi P, Carta A, Gavini E. (2014). Improvement of thymol properties by complexation with cyclodextrins: *in vitro* and *in vivo* studies. *Carbohydrate Polymers*, 102, pp. 393–399.
190. Mallardo S, De Vito V, Malinconico M, Volpe MG, Santagata G, Di Lorenzo ML. (2016). Poly(butylene succinate)-based composites containing β-cyclodextrin/d-limonene inclusion complex. *European Polymer Journal*, 79, pp. 82–96.
191. Abbaszadegan S, Al-Marzouqi AH, Salem AA, Amin A. (2015). Physicochemical characterizations of safranal-β-cyclodextrin inclusion complexes prepared by supercritical carbon dioxide and conventional methods. *Journal of Inclusion Phenomena and Macrocyclic Chemistry*, 83(3), pp. 215–226.
192. Ceborska M, Asztemborska M, Lipkowski J. (2012). Rare 'head-to-tail' arrangement of guest molecules in the inclusion complexes of (+)- and (–)-menthol with β-cyclodextrin. *Chemical Physics Letters*, 553, pp. 64–67.

7. Applications in the Textile Industry

Jin Xu

*Key Laboratory of Eco-textiles of Ministry of Education,
College of Textiles and Clothing,
Jiangnan University, Wuxi 214122, China*

7.1 Introduction

Cyclodextrins (CDs) have been used as auxiliaries in dyeing processes since the early 1990s [1, 2] and it soon became apparent that permanent fixation of CDs on the surface of fibers resulted in textiles with new properties. The first patent describing this technology was filed [3] in 1990, and detailed results were published shortly afterwards [4, 5].

CDs form host–guest complexes with a wide variety of solid, liquid, and gaseous compounds, including surfactants, dyes, perfumes, and antimicrobial agents. In the textile field, CDs are used to absorb unpleasant odors and also to complex and release fragrances or "skin-care-active" substances, such as vitamins, caffeine and menthol, as well as bioactive substances, such as biocides and insecticides [6]. CDs are also widely used to modify fiber surfaces, to remove surfactants from washed textiles, to control textile dyeing and to finish textiles [7]. Textile materials treated with CDs can be used as selective filters ("textile nanosponges") to adsorb small pollutants from wastewaters [6]. This chapter focuses on applications of CDs in the textile industry.

There are two methods for using CDs in the textile industry: the CD can be associated with the textile material without chemical or physical interactions, or the CD can be permanently fixed onto the fiber surface. In the latter case, reactive derivatives of CDs can be attached using nucleophilic groups in the textile, or unfunctionalized CDs can be cross-linked with different polymeric materials.

7.2 Textile dyeing applications

Leveling agents help to achieve uniform dyeing by slowing down dye exhaustion and by uniformly dispersing dye taken up by the fibers. CDs have been used as levelling agents for dyeing fibers such as cellulose [8] and polyamides [9] because they slow the dyeing process by forming complexes with the dyes. The complex of CD and dye disperses more slowly than the dye itself, and the dye is released and can be fixed to the fiber at higher temperatures.

Shibusawa *et al.* [10] reported that six azo disperse dyes (4-amino-4′-nitroazobenzene derivatives) formed inclusion complexes with α-, β- and γ-CDs. A computer simulation was used to assess the retarding action of β-CD on the dyeing rate of an azo disperse dye on secondary acetate fibers at 90°C. Improvements of color uniformity between fourfold and tenfold were achieved when PA66 and microfiber PA6 were dyed in the presence of CDs [7]. β-CD has been used in direct dyeing of cellulosic fabrics and in rinsing processes for direct dyed fabrics. The leveling effect of β-CD has been investigated using an exhaust dyeing method [8]. It has been reported that α- and β-CD can form inclusion complexes with sulfonated azo dyes (orange II, ponceau SX, allura red AC, and tartrazine) [11]. Because of high substantivity and rapid uptake over a small temperature range above the glass-transition temperature (T_g) of the fiber, cationic dyes have very low ability to migrate on polyacrylonitrile (PAN) fibers. In this case, uniformity of color can be improved by the use of β-CD as a dye retardant. Voncina *et al.* [7] reported that significant improvements in color uniformity and some improvements in color depth were achieved when PAN fibers were dyed in the presence of β-CD rather than a commercial retardant. Carpignano *et al.* [12] reported that using β-CD as a substitute for a commonly used commercial surfactant in the dyeing of polyester with disperse dyes produced satisfactory results and reduced the environmental impact of the exhausted dyebaths. Surface modification using β-CD and citric acid has also been used to improve the sharpness and color yield in inkjet printing of polyester fabrics using water-based pigments [13].

Ibrahim *et al.* [14] successfully used monochlorotriazine β-CD (MCT-β-CD) to modify the structure of woolen fabric and demonstrated that the modification improved printing properties using different dyestuffs. MCT-β-CD has also been used to improve the dyeability of flax fiber substrates [15]. The dyeing of wool fibers that had been pretreated with β-CD or MCT-β-CD was accompanied by formation of noncovalent host–guest inclusion complexes with components of the dye mix, and the coloration properties were thus enhanced [16]. The environmental impact of the dyeing process was also reduced by replacing more hazardous chemicals with nonpolluting CDs.

7.3 Textile washing processes

If surfactants are used during any step in textile processing, a small amount of the surfactant will remain adsorbed on the surface of the fibers and alter the dyeability, wettability, and absorptivity of the textile surface. The presence of adsorbed surfactants on the surface of the fibers also reduces the quality of hydrophobic finishing.

CDs and their derivatives are able to form complexes with surfactants in aqueous solution and thus remove most of the adsorbed surfactants from textile fibers [7]. CDs can also act as defoaming agents and thereby reduce water consumption. In a common laundering process, in which the wash cycle was followed by two rinse cycles with cold soft water and one rinse cycle with cold hard water, each lasting 3 min, addition of CD (3 g/L) to the final rinse water reduced the amount of residual surfactant on the laundered fabric from 209 to 134 ppm [17, 18].

Because CDs form complexes with perfumes or fragrances that are stable over long periods of time, they can be used in surfactants and other easy care products for textiles [19, 20]. The evaporation of perfumes from detergent powders is strongly inhibited by clathration with CDs. In aqueous solution, the clathrated perfumes are displaced from the CD cavity by surfactant molecules. The extent of this displacement varies largely with the chemical nature and/or steric factors of the competing molecules. In a study by Koch [21], the perfumes under investigation were released most easily by anionic surfactants. Under normal storage conditions, the complexation of perfumes by CDs provides good stabilization when mixed with detergent powder and almost quantitative release upon contact with water.

7.4 Functional finishing of textiles

7.4.1 *Fixation of CDs onto the textile*

There are currently two possible approaches for bonding CDs onto textile fibers: chemical bonding of modified CDs onto fiber surfaces [22] and covalent linking of CDs onto fiber surfaces using bi- or poly-functional reactive compounds [1]. Feasible interactions between β-CD and different textile fibers are shown in Table 7.1. Permanent grafting of CDs onto textile fibers allows the properties of the CDs to become intrinsic to the modified fibers and has led to a new generation of intelligent textiles with enhanced capacities for sorption and delivery of active molecules [23]. Covalent bonding of CDs onto textile fibers was first patented in 1980 by Szejtli [24], who described bonding CD onto alkali-swollen cellulose fibers using the cross-linking reagent epichlorohydrin.

Table 7.1.   Feasible interactions between β-CD and different textile fibers.

Parameter	Cotton	Wool	PES	PA	PAN	PP
Ionic interactions	−	+	−	+	+	−
Covalent bonds	+	+	−	+	−	−
van der Waals forces	−	−	+	+	+	+
Cross-linking agents	+	+	+	−	−	−
Graft polymerisation	+	+	+	+	+	+

Note: +, possible; −, not possible; PES, polyester; PA, polyamide; PAN, Polyacrylonitrile; PP, polypropylene.

Denter and Schollmeyer [25] demonstrated that functionalized CD derivatives (monochlorotriazinyl-β-CD, dihydroxypropyl-ethylhexylglycidyl-β-CD, hydroxypropyl-butylglycidyl-β-CD, hydroxypropyl-hydroxyhexyl-β-CD, hydroxypropyl-phenylglycidyl-β-CD, *o*-cresylglycidyl-β-CD, ethylhexylglycidyl-β-CD, hydroxypropyl-trimethylammoniumchloride-β-CD) can be fixed permanently onto polymer surfaces (cotton, regenerated cellulose, polyacrylonitrile, polyethylene terephthalate (PET), and PA66) using conventional textile-processing techniques. The CD cavities retain their ability to form inclusion complexes and special effects can be achieved by forming complexes with specific chemical substances. Wang [26] used 1,2,3,4-butanetetracarboxylic acid (BTCA) to anchor β-CD onto cellulosic fibers for depositing fragrance.

Ishifune *et al.* [27] developed a method for surface modification of carbon fibers using electro-oxidation and electroreduction and have successfully prepared a functional carbon fiber electrode bearing β-CD substituents.

Voncina and Marechal [28] cross-linked β-CD molecules onto the hydroxyl groups of cellulose using BTCA to form textiles with new functional properties.

Scalia *et al.* [29] attached β-CD to the cellulosic fabric, Tencel, using MCT-β-CD and then incorporated the sunscreen octyl methoxycinnamate into the β-CD cavities covalently bound to the cloth fibers.

Wang and Chen [30] were able to reduce the loss of volatile fragrance substances by forming inclusion complexes with β-CD molecules that had been anchored to cotton fabrics using a β-CD sol–gel solution method, involving 3-glycidyloxypropyl-trimethoxysilane, tetraethoxysilane, catalyzer, and solvent.

Ghoul *et al.* [31] attached CD molecules to polyamide fibers using citric acid as the cross-linking agent. The cross-linked polymer of CD and citric acid, which adhered physically to the fibers and was resistant to hot water washings, imparted new physical properties to the polyamide, and improved its affinity for dyes.

Ghemati *et al.* [32] grafted β-CD onto cellulose fibers and then included benzoic acid in the grafted cavities as a probe chemical. Finishing with quaternary ammonium salts produced fibers with attractive swelling behavior, dye adsorption, and antibacterial activity.

Park *et al.* [33] used a photografting technique to attach CDs to cellulose diacetate fibers, which are used mainly as cigarette filters. As well as removing harmful substances present in tobacco smoke, the modified fibers might have wider applications, including biomedical applications.

Voncina *et al.* [34] grafted β-CD onto PET textile materials using BTCA. The modified materials could be used to carry aromatherapy fragrances in products such as underwear and bed sheets or to absorb unpleasant smells, such as cigarette smoke.

Zhao *et al.* [35] reported a surface modification method for cellulose fiber that was based on supramolecular assembly. β-CD was covalently grafted onto the fiber surface, and then poly(epsilon-caprolactone) oligomers that had both ends capped with adamantane motifs were immobilized onto the cellulose fiber surface through host–guest inclusion complexation between β-CD and the adamantane motifs.

Nazi *et al.* [36] used response surface methodology to optimize the conditions for permanently fixing β-CD to cellulosic materials using an itaconic acid derivative.

Electrospun fibers based on aliphatic polymers have many potential biomedical applications, but their use is limited by poor wettability and lack of functional groups for attaching bioactive substances. Zhan *et al.* [37] designed multifunctional electrospun nanofibers based on an inclusion complex of poly(epsilon-caprolactone)-α-CD, which provided both structural support and multiple functionalities for further conjugation of bioactive components.

Kayaci *et al.* [38] attached three different types of native CD (α-CD, β-CD, and γ-CD) to electrospun polyester nanofibers by cross-linking with citric acid. The modified nanofibers removed polycyclic aromatic hydrocarbons from aqueous solution more efficiently than pristine polyester nanofibers, suggesting that the modified nanofibers might be useful as a filter material for water purification and waste treatment.

Medronho *et al.* [39] have revisited a simple synthetic procedure and examined the parameters. It was possible to fix CDs on cellulose using a simple polycarboxylic acid as the cross-linking agent.

Dehabadi *et al.* [40] investigated the use of polyaminocarboxylic acids as novel cross-linking agents for the fixation of β-CD on cotton fabrics and used the weight increase of treated samples to quantify fixation of β-CD.

Dong *et al.* [41] prepared antibacterial modified cellulose fiber by covalently bonding β-CD with cellulose fiber using citric acid as cross-linking agent, followed by inclusion of the antibiotic, ciprofloxacin hydrochloride.

Ammar *et al.* [42] found that polyacrylic acid was superior to citric acid or BTCA as a cross-linking agent for the functionalization of polypropylene fibers with CDs. The cross-linked polymer formed between polyacrylic acid and CD physically adhered to the fiber network and was resistant to hot water washing. Functionalization did not affect the mechanical properties of the native fabric.

Forouharshad *et al.* [43] prepared electrospun stereocomplex polylactide-based fibers functionalized with polyhedral oligomeric silsesquioxanes. These fibers then underwent a grafting reaction with β-CD molecules that were activated to nucleophilic substitution by monotosylation. The resulting fibers showed increased surface wettability and could find application in the removal of water pollutants.

7.4.2 *Functional-finishing applications*

Chemical finishing provides textiles with new functionalities and also makes them suitable for special applications. In the textile industry, there is increasing interest in the production of new and versatile fabrics with specific properties, including good feel and drape, flame retardancy, water and oil resistance, shrink resistance, antimicrobial properties, and optical effects. CDs have many potential applications in the functional finishing of textiles [7] since they can be used for fragrance release [44]; odor adsorption [45]; controlled release of antibacterial, fungicidal, or insect repellent compounds [7]; UV protection [46]; and stabilization of active ingredients.

Lee *et al.* [47] used N-methylol-acrylamide to synthesize a CD-containing monomer, which was then grafted onto cellulose fibers. The possibility of textile finishing of CD-containing cotton fibers was investigated using benzoic acid as an antibacterial finishing agent or vanillin as an aroma-finishing agent. The antibacterial activity of benzoic acid-treated samples was retained even after 10 laundering cycles and the vanillin fragrance lasted much longer than that of a control sample.

Lo *et al.* [48] permanently grafted β-CD onto the surface of the cellulosic fabric and then loaded benzoic acid, vanillin, iodine, N,N-diethyl-*m*-toluamide, and dimethyl phthalate onto the modified fabric. The included compounds were efficiently hosted in the CD cavities, and the surface properties of the fabric were not significantly modified by the chemical treatments.

Martel *et al.* [44] impregnated cotton, wool, and polyester fabrics finished with CDs with six different fragrance molecules and citronella oil. The samples retained the fragrance for 1 year, compared with only a few weeks for untreated fabrics. The efficiency of different CDs was in the order γ-CD > β-CD > α-CD, and the durability of the fragrance was directly proportional to the amount of CD grafted onto the fabric.

Fabric resiliency and ultraviolet-protection properties of cotton-containing fabrics were enhanced by ester cross-linking using polycarboxylic acids, both in the absence or presence of additives (chitosan, β-CD or choline chloride), followed by posttreatment with solutions of metal salt [49].

Scalia *et al.* [29] grafted β-CD molecules onto Tencel using MCT-β-CD and compared the ability of the treated fabric to retain sunscreen with that of untreated fabric. Both modified and unmodified fabrics were soaked in water–methanol mixtures containing 2% (v/v) of octyl methoxycinnamate and then subjected to several washing cycles. The treated fabric retained approximately eightfold more sunscreen than the untreated fabric and also showed enhanced photoprotective properties.

Gawish *et al.* [50] used oxidative atmospheric pressure plasma to activate the surface of PA66 fabrics, followed by graft copolymerization of glycidyl methacrylate and further reaction with triethylene tetramine, quaternary ammonium chitosan, or β-CD. The inner CD cavity was complexed with insecticidal perfumes, and the modified fabric was found to have efficient antimicrobial and insect repelling properties.

Kim *et al.* [51] prepared a β-CD derivative containing a cyanuric chloride (CC) moiety (β-CD-CC) by reaction of β-CD with CC under alkaline condition. The β-CD-CC was then used as a host material for inclusion complexes with various guest molecules. As an example, vanillin, a flavoring agent used in foods and other applications, was included in the β-CD-CC to make a perfume inclusion complex. UV analysis showed that the vanillin-included complex of β-CD-CC released vanillin more slowly than vanillin itself, suggesting that β-CD-CC might be suitable as a fragrance-finishing agent for textiles.

Abdel-Mohdy *et al.* [52] permanently fixed MCT-β-CD onto the surface of cotton fabrics and then formed inclusion complexes with the insecticides cypermethrin and prallethrin. The MCT-β-CD-finished cotton fabrics loaded with insecticides provided effective personal protection against mosquitoes and were effective in reducing malaria morbidity and mortality.

El Ghoul *et al.* [53] obtained inguinal meshes with improved antibiotic delivery properties by using CDs as finishing agents for polyamide fibers. The finishing process, which involved polymerization between citric acid and CDs, yielded a cross-linked polymer that physically adhered to the surface of the fibers. The fibers were functionalized with a hydroxypropylated derivate of γ-CD and charged with ciprofloxacin. The *in vitro* antibacterial activity of the modified fibers against *Staphylococcus aureus*, *Staphylococcus epidermidis*, and *Escherichia coli* was greatly superior to that of untreated fibers in a 24 h batch experiment in human blood plasma medium.

Wang *et al.* [54] investigated incorporation of the antibacterial agent miconazole nitrate into CD cavities covalently bonded onto cloth fibers. The cellulosic fabric was grafted with β-CD molecules through reaction with MCT-β-CD. The level of miconazole nitrate entrapped in the textile functionalized with MCT-β-CD was approximately eightfold higher than in the unmodified fabric. Antibacterial activity, measured by the shaker flask method, was markedly enhanced by impregnation of the MCT-β-CD-grafted textile with miconazole nitrate. The finished fabric retained more than 70% of its antibacterial activity even after ten washes whereas the antibacterial activity of the unmodified textile was almost entirely lost.

Zhang *et al.* [55] reported that the use of β-CD to form inclusion complexes during the fragrance finishing of cotton fabrics reduced the vapor pressure of the volatile fragrance oils and increased fragrance retention. The inclusion complexes were prepared by the immersion method under orthogonal experimental conditions, and the inclusion complex with lavender oil was found to have a unique molecular structure and inclusion nature. The lavender oil absorbed by β-CD was isolated by Soxhlet extraction, and the quantity was determined by UV–VIS spectrometry. The optimized inclusion conditions were found to be inclusion time, 2 h; temperature, 30°C; pH value, 9; and ratio of alcohol/water, 1:1.

Ibrahim and El-Zairy [46] reported that pretreatment of wool–polyester-blend fabric with MCT-β-CD modified the wool component to be able to form "host–guest" inclusion complexes with disperse dyes during the subsequent disperse printing step, thereby leading to union disperse printing. The treatment conditions were optimized, and the surface morphology was studied using scanning electron microscopy. Fixing of MCT-β-CD onto and/or within the wool component modified its structure and thereby increased its ability to pick up, adsorb, and retain the guest disperse dye vapors into its grafted hydrophobic cavities and enabled union-disperse printing with deeper shades and remarkable fastness properties.

Abdel-Halim *et al.* [56] used linear electron beam radiation to induce grafting of a glycidyl methacrylate/β-CD mixture onto cotton fabric. Chlorohexidin diacetate was incorporated into the cavities of CD fixed on the cotton fabric to form inclusion complexes with antimicrobial activity. Grafted fabrics loaded with chlorohexidin diacetate showed good antimicrobial activity compared with control and grafted fabrics that are not loaded with antimicrobial agent. The grafted fabrics loaded with chlorohexidin diacetate also exhibited good antimicrobial activity after five washes, and this lasting antimicrobial activity was attributed to the inclusion complex formed between chlorohexidin diacetate molecules and the cavities of CD.

El Shafei *et al.* [57] synthesized reactive polymers comprising MCT-β-CD grafted with butyl acrylate and investigated their application to cotton fabrics, along with epichlorohydrin and/or ZnO nanoparticles, using the conventional pad-dry-cure method. Grafting of this reactive preformed polymer onto cotton fabrics in the presence of ZnO nanoparticles and/or epichlorohydrin imparted antibacterial activity, 70% of which was retained after twenty washes. The grafted fabrics had improved air permeability, indicating that garments produced using these fabrics will have better breathability and comfortability.

Ammayappan *et al.* [58] reported that pretreatment of yarn-blended wool–cotton fabric with either a cellulase or a protease and subsequent finishing with a combination of Synthappret-BAP and β-CD imparted antishrink, antimicrobial, softening, and anticrease properties, depending on the type of finishing combination applied. The performance of the finished fabric depended on the type of finishing combination applied rather than the type of enzyme pretreatment. Pretreatment with savinase followed by finishing with a combination of Synthappret and Ceraperm-MW imparted both antishrink properties and softness, whereas pretreatment with bactosol followed by finishing with a combination of β-CD and sanitizer imparted antimicrobial activity and antishrink properties to the wool–cotton-blend fabric.

Nazi *et al.* [59] modified β-CD using itaconic acid, which contains both carboxyl and vinyl groups. This bifunctional compound was attached to β-CD via an esterification reaction, leaving the vinyl group available to perform free-radical polymerization reactions. After grafting onto cotton fibers, the presence of anchored CD nanoparticles on the surface of the fibers was demonstrated using scanning electron microscopy and Fourier transform infrared spectroscopy as well as by the ability of the attached CDs to form inclusion complexes.

Rukmani and Sundrarajan [60] produced antibacterial textiles with improved efficacy and durability by incorporating the monoterpene thymol into β-CD-grafted organic cotton. Grafting of β-CD onto the fabric was carried out using citric acid as cross-linker in the presence of sodium hypophosphite. Using the agar diffusion method, thymol-loaded ungrafted and grafted fabrics were shown to inhibit the growth of *Escherichia coli* more strongly than the growth of *Staphylococcus aureus*. The inhibitory effect was enhanced by inclusion of the antibacterial agent into grafted fabric. The antibacterial effect of the grafted material was maintained after 10 wash cycles whereas it was not retained in ungrafted fabric. Grafting of CD onto fabric and inclusion of thymol into the cavity thus produced a durable antibacterial textile.

Volatility of fragrance compounds can be reduced by molecular encapsulation and Khajavi *et al.* [61] used β-CD to host lavender essential oil on cotton fabric. Dimethyl dihydroxy ethylene urea was used for bonding the β-CD to the cotton fabric, and lavender essential oil then bonded to β-CD because of their mutual hydrophobic character. The fragrance of samples with molecularly encapsulated lavender essential oil was more durable than that of untreated or dimethyl dihydroxy ethylene urea-treated samples.

Using a process known as biopolishing, Sundrarajan and Rukmani [62] pretreated organic cotton with cellulase to improve the grafting yield of MCT-β-CD. Incorporation of the antibacterial compound thymol into the biopolished MCT-β-CD-grafted organic cotton produced a material with enhanced antibacterial properties and improved durability after repeated washing.

Abdel-Halim *et al.* [63] grafted β-CD onto cotton fabric by cross-linking with BTCA in the presence of sodium hypophosphite monohydrate as catalyst and then loaded the finished cotton fabric with the antimicrobial agent octenidine dihydrochloride. The finished cotton fabric retained a reasonable amount of antimicrobial activity, even after 20 wash cycles. The long-lasting antimicrobial activity was attributed to the hosting ability of the cavities present in CD moieties, which host the antimicrobial molecules and release them gradually.

Hebeish *et al.* [64] prepared cotton textiles with activity against blood-sucking insects, such as mosquitoes, by incorporating the insecticides permethrin and bioallethrin into the macromolecular structure of modified cotton fabrics. The cotton was chemically modified by grafting with glycidyl methacrylate alone, or in combination with β-CD, using fast electron beam irradiation. The modified cotton could be re-treated to increase the amount of attached CD and hence the number of cavities. The finished cotton fabrics were fast acting against mosquitoes and the amount of insecticide in the finished fabric was shown to increase with increasing amounts of fixed CD, which increased the number of cavities able to host the insecticides.

Kettel *et al.* [65] have developed nanogels with high levels of covalently embedded β-CD as a new and versatile method for coating textiles to protect against insects. The complexation potential of covalently embedded β-CD in nanogels was demonstrated using the widely used insecticide permethrin in aqueous environment. Permethrin-containing β-CD nanogels could be applied easily, homogeneously, and safely on keratin fibers, such as wool fabrics and human hairs. The permethrin concentration on the fibers was directly controlled by the permethrin content in the nanogels. Permethrin complexed in nanogels was removed from the textile during washing, but complexation of permethrin by β-CD domains in the nanogels protected the active ingredient from UV degradation. Bioassays using the larvae of *Tineola bisselliella* (clothes moth) and *Anthrenocerus australis* (carpet beetle) showed that the activity of permethrin was not reduced by complexation in β-CD gels and that treatment protected the wool fibers against degradation by the insect larvae.

Ghosha and Chipot [66] attached β-CD to 100% cotton and 50:50 cotton–polyester fabrics using sol-gel, prepared from 3-glycidoxypropyltrimethoxysilane and tetraethoxy orthosilicate, which attaches to cotton cellulose through covalent bonding. Aromatherapy essential oils, with eucalyptus, lavender, and lemon fragrances, were introduced into the β-CD cavities and the durability of the fragrances on the fabrics was rated by four judges for 6 weeks and up to six washings. The judges rated the fragrance intensity once a week during this period and, although the scent ratings decreased over 6 weeks and six washings, the fabrics retained some fragrance after 6 weeks.

7.5 Other applications of CDs in textile manufacturing

Actively moving polymers (AMPs) are stimuli-sensitive polymeric materials that can respond to specific changes in their environment by adopting certain macroscopic shapes. AMPs are used in textile applications such as fiber spinning, fabric manufacturing, and shape memory finishing technologies [67]. At the molecular level, AMPs are elastic polymer networks that consist of switches and netpoints. The netpoints determine the permanent shape of the polymer network and can be of a chemical (covalent bonds) or physical (noncovalent bonds) nature. Supramolecular complexes based on CD inclusions were reported to have physical netpoints [7].

Stimuli-responsive polymers (SRPs) are smart materials that show noticeable changes in properties in response to changes in environmental stimuli. SRPs have been used in textiles to improve or achieve textiles with smart functionalities [68]. One example of an SRP is thermal-responsive polymeric hydrogel (TRPG), which is a three-dimensional macromolecular gel network that contains a large fraction of water within its structure [69, 70]. TRPG with incorporated β-CD molecules can further enhance the controlled-release properties of hydrogel-modified textiles [71,72].

Ducoroy *et al.* [73] evaluated the capacity of new cationic exchange textiles (CETs) to take up four heavy metals (Pb^{2+}, Cd^{2+}, Zn^{2+}, and Ni^{2+}) from water. The CETs are nonwoven PETs, coated with a copolymer (polyester) of β-CD and citric acid (CTR), 1,2,3,4–butanetetracarboxylic acid (BTCA), or polyacrylic acid. Investigation of the effect of different experimental conditions (time, pH, and metal mixture) on the efficiencies of these three CETs (in both carboxylic and carboxylate forms) and the uncoated (β-CD-citric acid) copolymer showed that CETs coated with CTR or BTCA were most effective for depollution applications, in terms of the metal adsorption/cost ratio.

Grigoriu *et al.* [74] have developed a method for grafting MCT-β-CD onto bast fibers during wet spinning and have used this method to prepare a textile material with new functional properties for medical applications. Several simultaneous mechanical and chemical processing trials were carried out to optimize the process parameters, and mathematical methods of dispersion analysis and regression were also applied.

"Cosmetotextiles" merge cosmetics or pharmaceuticals with textiles through the process of microencapsulation. One application is the treatment of chronic venous insufficiency in legs using elastic bandages loaded with natural products that possess phlebotonic properties. Cravotto *et al.* [75] have developed an efficient synthetic procedure for the preparation of β-CD-grafted viscose using a two-step ultrasound-assisted reaction together with a suitable cosmetic preparation, containing natural substances and extracts (aescin, menthol, *Centella asiatica*, and *Ginkgo biloba*), to recharge the CD-grafted textile. The efficacy of the new cosmetotextile was demonstrated by *in vitro* studies of diffusion through membranes, cutaneous permeation, and accumulation in porcine skin. Aescin was used as a reference compound, and its concentration in the different compartments was monitored by high-performance liquid chromatography (HPLC) analysis. The new cosmetotextile was cost effective, showed excellent application compliance, and was easily recharged, making it suitable for possible industrial production.

Another potential application of cosmetotextiles is wound healing. Nada *et al.* [76] have grafted per-(2,3,6 O-allyl)-β CD onto cotton fibers using plasma and conventional thermofixation methods. The cotton was activated by surface modification of the fibers with iododeoxycellulose, cellulose peroxide, and a cellulose diazonium salt. Several plasma machines and conventional thermal techniques for fixation were studied. Iododeoxycellulose gave the best results using thermofixation with the *in situ* mode of the APJeT atmospheric plasma machine. Linoleic, ricinoleic, and oleic acids were included in the grafted CD fabrics as possible wound-healing agents. The treated samples, both with and without included fatty acids, showed promising cytocompatibility.

Karim *et al.* [77] have developed a system based on immobilized horseradish peroxidase (HRP) for the removal of azo dyes from textile effluents. The maximum binding of HRP on a β-CD–chitosan complex was achieved at pH 8.0. Oxidative removal of azo dyes from effluent was monitored by HPLC analysis, and the cross-linked HRP–β-CD–chitosan complex was found to be more active, effective, and stable than uncross-linked or water-soluble enzyme preparations.

Amiri *et al.* [78] grafted a temperature-sensitive hydrogel, with the capability of forming inclusion complexes with guest molecules, onto the surface of nonwoven polypropylene using N-isopropylacrylamide monomer and acrylamidomethyl-β-CD. The grafted hydrogel maintained its temperature sensitivity and could have applications in both textile and pharmaceutical industries.

As part of a project funded by the European Union, Semeraro *et al.* [79] have investigated the ability of different commercially available CDs to remove industrial textile dyes from wastewater. The main aim of the project is to validate CDs as encapsulation agents for textile dyes from wastewater and to recover and reuse both clean water and dyes for other industrial processes.

The discharge of metals and dyes poses a serious threat to public health and the environment. What is worse, these two hazardous pollutants are often found together in industrial wastewaters, making treatment more challenging. To address this problem, Zhao *et al.* [80] have developed an EDTA-cross-linked β-CD bifunctional adsorbent for simultaneous adsorption of metals and dyes. The adsorbent is fabricated through the polycondensation reaction of β-CD with EDTA as a cross-linker. The CD cavities are expected to capture dye molecules through the formation of inclusion complexes, and the EDTA units act as adsorption sites for metals. Its facile and green fabrication, efficient sorption performance, and excellent reusability indicate that EDTA-β-CD has potential for practical applications in integrative and efficient treatment of coexistent toxic pollutants.

7.6 Conclusions

CDs play important roles in innovative textile processing and in the functionalization of textiles. Since 1980, when Szejtli first described the bonding of CDs onto textile fibers, much attention has been payed to the application of CDs onto textile substrates. The industrial application of CDs is now widespread, and new uses in the areas of medical and technical textiles are being explored. There are many possibilities for the development of new textile products with advanced properties based on CDs.

References

1. Denter U, Buschmann HJ, Schollmeyer E. (1991). Cyclodextrins and dextrins as new auxiliaries in dyeing. *Melliand Textilberichte*, 72, pp. 1012–1014.
2. Denter U, Buschmann HJ, Schollmeyer E. (1991). Trichromatic dyeing of cotton: interactions between dyes and auxiliaries. *Textilveredlung*, 26, pp. 113–116.
3. Poulakis KDC, Buschmann HJDR, Schollmeyer EPDR. (1990). Textiles material sowie verfahren zur herstellung eines derartigen textilen materials. Google Patents.
4. Ruppert S, Knittel D, Buschmann HJ, Wenz G, Schollmeyer E. (1997). Fixierung von β-cyclodextrinderivaten auf polyesterfasermaterial. *Starch-Stärke*, 49, pp. 160–164.
5. Szejtli J. (2003). Cyclodextrins in the textile industry. *Starch-Stärke*, 55, pp. 191–196.
6. Voncina B, Vivod V. (2013). Cyclodextrins in textile finishing. In: Günay M (Ed.), *Eco-Friendly Textile Dyeing and Finishing*. Rijeka: InTechOpen.
7. Vončina B, Vivod V, Jaušovec D. (2007). β-Cyclodextrin as retarding reagent in polyacrylonitrile dyeing. *Dyes and Pigments*, 74, pp. 642–646.
8. Cireli A, Yurdakul B. (2006). Application of cyclodextrin to the textile dyeing and washing processes. *Journal of Applied Polymer Science*, 100, pp. 208–218.
9. Shibusawa T, Hamayose T, Sasaki M. (1976). ChemInform Abstract: Dyeing properties of disperse dyes. VI. Interactions between β-cyclodextrin and 4-aminoazobenzene derivatives and the effect of the interactions on the dyeing properties of 4-aminoazobenzene derivatives. *Chemischer Informationsdienst*, 7.
10. Shibusawa T, Okamoto J, Abe K, Sakata K, Ito Y. (1998). Inclusion of azo disperse dyes by cyclodextrins at dyeing temperature. *Dyes and Pigments*, 36, pp. 79–91.
11. Zhang H, Chen G, Wang L, Ding L, Tian Y, Jin W, Zhang H. (2006). Study on the inclusion complexes of cyclodextrin and sulphonated azo dyes by electrospray ionization mass spectrometry. *International Journal of Mass Spectrometry*, 252, pp. 1–10.
12. Carpignano R, Parlati S, Piccinini P, Savarino P, De Giorgi MR, Fochi R. (2010). Use of β-cyclodextrin in the dyeing of polyester with low environmental impact. *Coloration Technology*, 126, pp. 201–208.
13. Chen L, Wang C, Tian A, Wu M. (2012). An Attempt of improving polyester inkjet printing performance by surface modification using β-cyclodextrin. *Surface and Interface Analysis*, 44, pp. 1324–1330.

14. Ibrahim NA, Abdalla WA, El-Zairy EMR, Khalil HM. (2013). Utilization of monochloro-triazine β-cyclodextrin for enhancing printability and functionality of wool. *Carbohydrate Polymers*, 92, pp. 1520–1529.
15. Coman D, Oancea S, Vrinceanu N, Stoia M. (2014). Sonication and conventional dyeing procedures of flax fibres with allium cepa anthocyanin extract. *Cellulose Chemistry and Technology*, 48, pp. 145–157.
16. Ibrahim NA, Khalil HM, Eid BM. (2015). A cleaner production of ultra-violet shielding wool prints. *Journal of Cleaner Production*, 92, pp. 187–195.
17. Michael D, Chris S, Thomas M, Frank E. (1998). Washing or laundering post-treatment agents, WO9813456.
18. Dieter D. (2000). Fabric rinse aid, useful in home and commercial laundry and dry cleaning, adjusts final rinse bath to skin-neutral pH and eliminates environmental contamination and detergent residue, DE19923303.
19. Bacon DR, Toan T. (1996). Detergent compositions containing enduring perfume, US5500154.
20. Gardlik JM, Trinh T, Banks TJ, Benvegnu F. (1990). Treatment of fabric with perfume/cyclodextrin complexes, EP0392606.
21. Koch J. (1982). Stabilisation and controlled release of perfume in detergents. In: Szejtli J (Ed.), *Proceedings of the First International Symposium on Cyclodextrins: Budapest, Hungary*. Dordrecht: Springer, pp. 487–496.
22. Issazadeh-Baltorki H, Khoddami A. (2014). Cyclodextrin-coated denim fabrics as novel carriers for ingredient deliveries to the skin. *Carbohydrate Polymers*, 110, pp. 513–517.
23. Knittel D, Schollmeyer E. (2000). Technologies for a new century. Surface modification of fibres. *Journal of the Textile Institute*, 91, pp. 151–165.
24. Szejtli J, Zsadon B, Fenyvesi E, Horvath N, Tudos F. (1982). Sorbents of cellulose basis capable of forming inclusion complexes and a process for the preparation thereof, US Patent 4357468.
25. Denter U, Schollmeyer E. (1996). Surface modification of synthetic and natural fibres by fixation of cyclodextrin derivatives. *Journal of Inclusion Phenomena and Molecular Recognition in Chemistry*, 25, pp. 197–202.
26. Wang CX, Chen SL. (2004). Surface modification of cotton fabrics with β-cyclodextrin to impart host-guest effect for depositing fragrance. *AATCC Review*, 4, pp. 25–28.
27. Ishifune M, Suzuki R, Mima Y, Uchida K, Yamashita N, Kashimura S. (2005). Novel electrochemical surface modification method of carbon fiber and its utilization to the preparation of functional electrode. *Electrochimica Acta*, 51, pp. 14–22.
28. Voncina B, Le Marechal AM. (2005). Grafting of cotton with β-cyclodextrin via poly(carboxylic acid). *Journal of Applied Polymer Science*, 96, pp. 1323–1328.
29. Scalia S, Tursilli R, Bianchi A, Lo Nostro P, Bocci E, Ridi F, Baglioni P. (2006). Incorporation of the sunscreen agent, octyl methoxycinnamate in a cellulosic fabric grafted with β-cyclodextrin. *International Journal of Pharmaceutics*, 308, pp. 155–159.
30. Wang CX, Chen SL. (2006). Surface treatment of cotton using β-cyclodextrins sol-gel method. *Applied Surface Science*, 252, pp. 6348–6352.

31. El Ghoul Y, Martel B, Morcellet M, Campagne C, El Achari A, Roudesli S. (2007). Mechanical and physico-chemical characterization of cyclodextrin finished polyamide fibers. *Journal of Inclusion Phenomena and Macrocyclic Chemistry*, 57, pp. 47–52.

32. Ghemati D, Oudia A, Aliouche D, Lamouri S. (2009). Biotreatment on cellulose fluff pulp: quaternary ammonium salts finish and grafting with β-cyclodextrin, *Applied Biochemistry and Biotechnology*, 159, pp. 532–544.

33. Park JY, Kong B, Chi YS, Kim Y-G, Choi IS. (2009). Aryl azide-based photografting of β-cyclodextrin onto cellulose diacetate fibers. *Bulletin of the Korean Chemical Society*, 30, pp. 1851–1854.

34. Voncina B, Vivod V, Chen WT. (2009). Surface modification of PET fibers with the use of β-cyclodextrin. *Journal of Applied Polymer Science*, 113, pp. 3891–3895.

35. Zhao Q, Wang S, Cheng X, Yam RCM, Kong D, Li RKY. (2010). Surface modification of cellulose fiber via supramolecular assembly of biodegradable polyesters by the aid of host-guest inclusion complexation. *Biomacromolecules*, 11, pp. 1364–1369.

36. Nazi M, Malek RMA, Moghadam MB. (2012). Effect of processing conditions on producing a reactive derivative from β-cyclodextrin with itaconic acid. *Starch-Stärke*, 64, pp. 794–802.

37. Zhan J, Singh A, Zhang Z, Huang L, Elisseeff JH. (2012). Multifunctional aliphatic polyester nanofibers for tissue engineering. *Biomatter*, 2, pp. 202–212.

38. Kayaci F, Aytac Z, Uyar T. (2013). Surface modification of electrospun polyester nanofibers with cyclodextrin polymer for the removal of phenanthrene from aqueous solution. *Journal of Hazardous Materials*, 261, pp. 286–294.

39. Medronho B, Andrade R, Vivod V, Ostlund A, Miguel MG, Lindman B, Voncina B, Valente AJM. (2013). Cyclodextrin-grafted cellulose: physico-chemical characterization. *Carbohydrate Polymers*, 93, pp. 324–330.

40. Dehabadi VA, Buschmann H-J, Gutmann JS. (2014). A novel approach for fixation of β-cyclodextrin on cotton fabrics. *Journal of Inclusion Phenomena and Macrocyclic Chemistry*, 79, pp. 459–464.

41. Dong C, Ye Y, Qian L, Zhao G, He B, Xiao H. (2014). Antibacterial modification of cellulose fibers by grafting β-cyclodextrin and inclusion with ciprofloxacin. *Cellulose*, 21, pp. 1921–1932.

42. Ammar C, El Ghoul Y, El Achari A. (2015). Finishing of polypropylene fibers with cyclodextrins and polyacrylic acid as a crosslinking agent. *Textile Research Journal*, 85, pp. 171–179.

43. Forouharshad M, Putti M, Basso A, Prato M, Monticelli O. (2015). Biobased system composed of electrospun Sc-PLA/POSS/cyclodextrin fibers to remove water pollutants. *ACS Sustainable Chemistry & Engineering*, 3, pp. 2917–2924.

44. Martel B, Morcellet M, Ruffin D, Vinet F, Weltrowski L. (2002). Capture and controlled release of fragrances by CD finished textiles. *Journal of Inclusion Phenomena and Macrocyclic Chemistry*, 44, pp. 439–442.

45. Sariisik M, Okur S, Asma S. (2012). Odor adsorption kinetics on modified textile materials using quartz microbalance technique. *Acta Physica Polonica A*, 121, pp. 243–246.

46. Ibrahim NA, El-Zairy EMR. (2009). Union disperse printing and UV-protecting of wool/polyester blend using a reactive β-cyclodextrin. *Carbohydrate Polymers*, 76, pp. 244–249.

47. Lee MH, Yoon KJ, Ko SW. (2000). Grafting onto cotton fiber with acrylamidomethylated β-cyclodextrin and its application. *Journal of Applied Polymer Science*, 78, pp. 1986–1991.

48. Lo Nostro P, Fratoni L, Baglioni P. (2002). Modification of a cellulosic fabric with β-cyclodextrin for textile finishing applications. *Journal of Inclusion Phenomena and Macrocyclic Chemistry*, 44, pp. 423–427.

49. Ibrahim NA, Refai R, Youssef MA, Ahmed AF. (2005). Proper finishing treatments for sun-protective cotton-containing fabrics. *Journal of Applied Polymer Science*, 97, pp. 1024–1032.

50. Gawish SM, Ramadan AM, Cornelius CE, Bourham MA, Matthews SR, McCord MG, Wafa DM, Breidt F. (2007). New functionalities of PA6, 6 fabric modified by atmospheric pressure plasma and grafted glycidyl methacrylate derivatives. *Textile Research Journal*, 77, pp. 92–104.

51. Ko JH, Park YC, Kim JW, Kim YH. (2007). Synthesis of β-cyclodextrin derivative containing cyanuric chloride moiety and its application to fragrant finish of textiles. *Textile Science and Engineering*, 44, pp. 183–186.

52. Abdel-Mohdy FA, Fouda MMG, Rehan MF, Aly AS. (2008). Repellency of controlled-release treated cotton fabrics based on cypermethrin and prallethrin. *Carbohydrate Polymers*, 73, pp. 92–97.

53. El Ghoul Y, Blanchemain N, Laurent T, Campagne C, El Achari A, Roudesli S, Morcellet M, Martel B, Hildebrand HF. (2008). Chemical, biological and microbiological evaluation of cyclodextrin finished polyamide inguinal meshes. *Acta Biomaterialia*, 4, pp. 1392–1400.

54. Wang J-h, Cai Z. (2008). Incorporation of the antibacterial agent, miconazole nitrate into a cellulosic fabric grafted with β-cyclodextrin. *Carbohydrate Polymers*, 72, pp. 695–700.

55. Yiyu Z, Lu W, Campagne C, Abdessemed R. (2008). Fragrant finishing of cotton fabric with lavender oil via β-cyclodextrin technology. *Journal of Textile Research*, 29, pp. 94–97.

56. Abdel-Halim ES, Fouda MMG, Hamdy I, Abdel-Mohdy FA, El-Sawy SM. (2010). Incorporation of chlorohexidin diacetate into cotton fabrics grafted with glycidyl methacrylate and cyclodextrin. *Carbohydrate Polymers*, 79, pp. 47–53.

57. El Shafei A, Shaarawy S, Hebeish A. (2010). Application of reactive cyclodextrin poly butyl acrylate preformed polymers containing nano-ZNO to cotton fabrics and their impact on fabric performance. *Carbohydrate Polymers*, 79, pp. 852–857.

58. Ammayappan L, Moses JJ, Senthil KA, Raja ASM, Jimmy LKC. (2011). Performance properties of multi-functional finishes on the enzyme-pretreated wool/cotton blend fabrics. *Textile Coloration and Finishing*, 23, pp. 1–10.

59. Nazi M, Malek RMA, Kotek R. (2012). Modification of β-cyclodextrin with itaconic acid and application of the new derivative to cotton fabrics. *Carbohydrate Polymers*, 88, pp. 950–958.

60. Rukmani A, Sundrarajan M. (2012). Inclusion of antibacterial agent thymol on β-cyclodextrin-grafted organic cotton. *Journal of Industrial Textiles*, 42, pp. 132–144.
61. Khajavi R, Ahrari M, Toliyat T, Bahadori L. (2013). Molecular encapsulation of lavender essential oil by β-cyclodextrin and dimethyl dihydroxy ethylene urea for fragrance finishing of cotton fabrics. *Asian Journal of Chemistry*, 25, pp. 459–465.
62. Sundrarajan M, Rukmani A. (2013). Biopolishing and cyclodextrin derivative grafting on cellulosic fabric for incorporation of antibacterial agent thymol. *Journal of the Textile Institute*, 104, pp. 188–196.
63. Abdel-Halim ES, Al-Deyab SS, Alfaifi AYA. (2014). Cotton fabric finished with β-cyclodextrin: Inclusion ability toward antimicrobial agent. *Carbohydrate Polymers*, 102, pp. 550–556.
64. Hebeish A, El-Sawy SM, Ragaei M, Hamdy IA, El-Bisi MK, Abdel-Mohdy FA. (2014). New textiles of biocidal activity by introduce insecticide in cotton-poly (GMA) copolymer containing β-Cd. *Carbohydrate Polymers*, 99, pp. 208–217.
65. Kettel MJ, Schaefer K, Groll J, Moeller M. (2014). Nanogels with high active β-cyclodextrin content as physical coating system with sustained release properties. *ACS Applied Materials & Interfaces*, 6, pp. 2300–2311.
66. Ghosha S, Chipot N. (2015). Embedding aromatherapy essential oils into textile fabric using β-cyclodextrin inclusion compound. *Indian Journal of Fibre & Textile Research*, 40, pp. 140–143.
67. Hu J, Chen S. (2010). A review of actively moving polymers in textile applications. *Journal of Materials Chemistry*, 20, pp. 3346–3355.
68. Jinlian H, Harper M, Guoqiang L, Samuel II. (2012). A review of stimuli-responsive polymers for smart textile applications. *Smart Materials and Structures*, 21, p. 053001.
69. Osada Y, Matsuda A. (1995). Shape memory in hydrogels. *Nature*, 376, p. 219.
70. Huang G, Gao J, Hu Z, John JVS, Ponder BC, Moro D. (2004). Controlled drug release from hydrogel nanoparticle networks. *Journal of Controlled Release*, 94, pp. 303–311.
71. Bhaskar M, Chandrasekar R, Radhakrishnan G. (2002). Separation of harmful chlorophenols by cyclodextrin-assisted capillary electrokinetic chromatography. *Journal of Separation Science*, 25, pp. 1143–1146.
72. Liu Y-Y, Fan X-D. (2002). Synthesis and characterization of pH- and temperature-sensitive hydrogel of N-isopropylacrylamide/cyclodextrin based copolymer. *Polymer*, 43, pp. 4997–5003.
73. Ducoroy L, Bacquet M, Martel B, Morcellet M. (2008). Removal of heavy metals from aqueous media by cation exchange nonwoven PET coated with β-cyclodextrin-polycarboxylic moieties. *Reactive and Functional Polymers*, 68, pp. 594–600.
74. Grigoriu A, Racu C, Diaconescu RM, Grigoriu A-M. (2010). Modelling of the simultaneous wet spinning-grafting process of hemp fibres meant for medical textiles. *Industria Textila*, 61, pp. 112–116.

75. Cravotto G, Beltramo L, Sapino S, Binello A, Carlotti ME. (2011). A new cyclodextrin-grafted viscose loaded with aescin formulations for a cosmeto-textile approach to chronic venous insufficiency. *Journal of Materials Science: Materials in Medicine*, 22, pp. 2387–2395.
76. Nada AA, Hauser P, Hudson SM. (2011). The grafting of Per-(2,3,6-O-allyl)-β cyclodextrin onto derivatized cotton cellulose via thermal and atmospheric plasma techniques. *Plasma Chemistry and Plasma Processing*, 31, pp. 605–621.
77. Karim Z, Adnan R, Husain Q. (2012). A β-Cyclodextrin-Chitosan Complex As The Immobilization Matrix For Horseradish Peroxidase And Its Application For The Removal Of Azo Dyes From Textile Effluent. *International Biodeterioration & Biodegradation*, 72, pp. 10–17.
78. Amiri S, Duroux L, Nielsen TT, Larsen KL. (2014). Preparation and characterization of a temperature-sensitive nonwoven poly(propylene) with increased affinity for guest molecules. *Journal of Applied Polymer Science*, 131. doi: 10.1002/app.40497.
79. Semeraro P, Rizzi V, Fini P, Matera S, Cosma P, Franco E, Garcia R, Ferrandiz M, Nunez E, Gabaldon JA, Fortea I, Perez E, Ferrandiz M. (2015). Interaction between industrial textile dyes and cyclodextrins. *Dyes and Pigments*, 119, pp. 84–94.
80. Zhao F, Repo E, Yin D, Meng Y, Jafari S, Sillanpaa M. (2015). EDTA-cross-linked β-cyclodextrin: an environmentally friendly bifunctional adsorbent for simultaneous adsorption of metals and cationic dyes. *Environmental Science & Technology*, 49, pp. 10570–10580.

8. Applications in Analytical Chemistry

Xuehong Li

School of Food and Bioengineering,
Zhengzhou University of Light Industry,
Zhengzhou 450002, China

8.1 Introduction

Cyclodextrins (CDs) are cyclic, nonreducing oligosaccharides consisting of -glucopyranose units bonded through α-1,4 linkages. The most abundant natural CDs are α-, β- and γ-CD, which have six, seven, and eight glucose units, respectively [1]. Native CDs are crystalline, nonhygroscopic and homogeneous compounds, which form torus-like macrocycles consisting of glucopyranose units. CD molecules have secondary hydroxyl groups at the 2- and 3-positions that are located on the upper, broad edge and primary hydroxyl groups at the 6-position that are located on the narrow side of the molecule. The torus-like geometry results in a hydrophilic exterior with attached hydroxyl groups and a relatively hydrophobic internal cavity, which allows CD molecules to easily form inclusion complexes with a wide variety of organic and inorganic molecules and noble gases. Complexation only takes place, however, when there is steric compatibility between the CD cavity and the guest molecule and when the affinity of the guest molecule for the CD cavity is higher than for the external solvent. Besides steric compatibility and hydrophobic interactions, both van der Waals interactive forces and hydrogen-bonding play important roles in the complexation process. A reasonable three-point interaction model has been developed to explain the stereoselectivity observed during inclusion complex formation by CDs. The steric matching, and consequently the energy of interaction, between the two enantiomers and the three points of attachment can result in the separation of positional isomers, functional groups, homologues, and enantiomers [2–4]. The CD scaffold also allows for assembly of functional groups with controlled geometry. CDs can be readily modified through substitution of the hydroxyl groups, giving rise to derivatives with significantly different properties, especially increased solubility and altered hydrophobicity of the cavity.

The properties of CDs and their derivatives allow their extensive utilization in analytical chemistry, which is almost exclusively related to this host–guest type molecular recognition process. CDs have been used effectively for sample preparation, chiral separations and to improve separation processes in many chromatographic techniques (as mobile phases, stationary phases, or stationary-phase additives). CDs are thermostable to reasonably high temperatures, which is important for gas chromatography (GC) and also stable over a very wide pH range (2–12). CDs do not absorb radiation in the region normally associated with UV detection (200–350 nm), facilitating their application in liquid chromatographic techniques. CDs can also be used as enzyme inhibitors in affinity techniques. In spectrometry, CDs can serve as complexing agents in ultraviolet/visible (UV/VIS) spectrophotometry, chiral shift agents in nuclear magnetic resonance (NMR) spectroscopy, and selective agents to alter spectra in circular dichroism analysis. As chiral recognition agents, CDs are of great importance in various electrophoresis techniques [5, 6]. CDs also have a wide range of application in microdialysis, solid- and liquid-phase extractions, separations through liquid and composite membranes and molecularly imprinted polymers.

8.2 Applications of CDs in sample preparation

Sample preparation is crucial in any analytical process. Usually, the concentration of the component of interest is too low for detection and sample preparation is used to concentrate the component to adequate levels for measurement. The selective molecular interactions of CD cavities with different dimensions provide the possibility for selective trapping and removal of either the target component or the interfering matrix components from complex analytical samples. Such processes improve the accuracy and reliability of most common analytical procedures by decreasing or eliminating matrix effects and have been employed to isolate or selectively enrich analytes from food samples, pharmaceutical products, tissue samples, and environmental samples [7, 8].

For example, CDs can be used to prepare CD-coated adsorbents that can selectively retain target analytes from dilute complex samples. The retained analyte is then eluted for analysis by adding a competitive complex-forming analytically inert agent. This competitive displacement principle has been successfully used for different analytes, including steroidal compounds, bile acids, and pharmaceuticals. CDs can also be used for the capture and recovery of target volatile organic solvents from gaseous matrices. CD-enabled silica particles can provide good reproducibility, with a coefficient of variation below 6% and high recovery rates [9].

CDs can also be used to remove interfering components from complex matrices by selective precipitation. For instance, CDs can remove lipids and lipoproteins from lipemic serum by complexing and precipitating these components. This CD-based serum clarification is simple, nonhazardous and effective. Pretreatment of samples with α-CD selectively removed fatty acids that cause interference during the automated colorimetric determination of serum calcium levels and eliminated the interference [10].

8.3 Applications of CDs in chromatographic separations

One of the most distinctive functional properties of CDs is the ability to form reversible, noncovalent inclusion complexes with compounds that geometrically fit inside the cavity. For chiral separation, a portion of the guest molecule must enter the hydrophobic cavity and a hydrogen bonding region of the guest molecule must interact with the edge of the cavity. The formation of inclusion complexes between CDs and guests is thus greatly affected by the hydrophobicity, shape, and size of guest molecules. CD complexation is highly selective and can provide highly selective systems for chromatographic separations. The partitioning and binding of hydrophobic and hydrophilic organic molecules to the CD cavity can be much more selective than the partitioning and binding to a single solvent or to a single traditional stationary phase. For this reason, CDs are especially useful in typically difficult chromatographic separations of enantiomers, diastereomers, structural isomers, and geometric isomers. Among the three types of CD, β-CD has found the widest application. There are some structural differences between the CDs (β-CD has a pronounced kink in its structure, whereas α-CD and γ-CD are more planar), which affect the type of molecules that can be complexed. α-CD can complex single phenyl groups or naphthyl groups end-on, β-CD can accept naphthyl groups and heavily substituted phenyl groups, and γ-CD is effective for complexing bulky steroid-type molecules [11].

8.3.1 *CDs in gas chromatography*

CDs and their derivatives can be used in stationary phases in GC for the separation of a wide variety of compounds. The overwhelming majority of applications are separations of optical and positional isomers and, in these cases, the elution order of the resolved compounds is in accordance with their binding affinity to the CD immobilized on the stationary resin [12]. Native CDs are rarely used in stationary phases because of their relatively low thermal stability and low separation efficiency. The high potential of CDs for enantioselective separations is due to the different molecular dimensions of the cavities, together with differences in the reactivity of the 2-, 3- and 6-hydroxy groups of the glucose moieties, which can be derivatized by regioselective alkylation and acylation. Much effort has gone into devising new synthetic routes to CD derivatives, and more than 50 have been so far designed and used successfully in GC separations. Almost 500 papers have described the benefits of short-chain peralkylated CDs. Methylation of the primary OH groups and substitution of the secondary OH groups with longer alkyl chains (e.g., *n*-butyl and *n*-pentyl) greatly improved enantioselectivity. Two of the most frequently used CD derivatives in capillary GC separations are heptakis-2,3-di-O-methyl-6-O-tert-butyl-dimethyl-silyl-β-CD (TBDMS) and 2,6-di-O-pentyl-3-O-trifluoro acetyl-β-CD (DPTABCD) [13]. These derivatives are oils and bear *t*-butyl and *n*-pentyl groups that allow improved distribution and orientation of the CDs in the polysiloxane carrier matrix. The use of derivatized α-, β- and γ-CD dissolved in polysiloxanes has a number of advantages: (1) the unique physicoselectivity of polysiloxanes is combined with the inherent chemoselectivity of CDs; (2) the universal coating properties of polysiloxanes lead to high resolution, high efficiency, and thermally stable (up to 250°C) capillary columns for high-resolution capillary GC; (3) high melting points or phase transitions of derivatized CDs are not detrimental to column performance and can be disregarded; and (4) multicomponent (mixed) CD-based stationary phases can be used [14]. Popular CD-derivatized chiral stationary phases (CD-CSPs) used for enantioselective separation by GC are listed in Table 8.1 [15].

Dissolution of derivatized CDs in semipolar polysiloxanes (e.g., OV-1701) represents the most useful approach for enantioseparation

Table 8.1. Derivatized cyclodextrins for enantioselective GC complementing the use of permethylated-β-cyclodextrin.

No.	Derivatized cyclodextrins
1	Hexakis(2,3,6-tri-O-pentyl)-α-cyclodextrin
2	Hexakis(3-O-acetyl-2,6-di-O-pentyl)-α-cyclodextrin
3	Hexakis(2,3,6-tri-O-pentyl)-β-cyclodextrin
4	Hexakis(3-O-acetyl-2,6-di-O-pentyl)-β-cyclodextrin
5	Octakis(2,3,6-tri-O-pentyl)-γ-cyclodextrin
6	Octakis(3-O-butanoyl-2,6-di-O-pentyl)-γ-cyclodextrin (Lipodex E)
7	Hexakis[(per-O-2-hydroxypropyl)-per-O-methyl]-α-cyclodextrin (PMHP-α-CD)
8	Hexakis[(per-O-2-hydroxypropyl)-per-O-methyl]-β-cyclodextrin (PMHP-β-CD)
9	Hexakis(2,6-di-O-pentyl)-α-cyclodextrin (dipentyl-α-CD)
10	Heptakis(2,6-di-O-pentyl)-β-cyclodextrin (dipentyl-β-CD)
11	Heptakis(3-O-trifluoroacetyl-2,6-di-O-pentyl)-β-cyclodextrin (DPTFA-β-CD)
12	Octakis(3-O-trifluoroacetyl-2,6-di-O-pentyl)-γ-cyclodextrin (DPTFA-γ-CD)
13	Heptakis(2,3-di-O-acetyl-6-O-tert-butyldimethylsily)-β-cyclodextrin
14	Heptakis(2,3-di-O-methyl-6-O-*tert*-butyldimethylsily)-β-cyclodextrin

by GC. One strategy that has been used to dissolve derivatized CDs in semipolar polysiloxanes is a dilution method that involves fixing the enantioselective CD selector to a poly(dimethylsiloxane) backbone by a permanent chemical linkage. This approach yielded a CSP composed of a chiral polysiloxane containing CD, which was termed Chirasil-β-Dex, by analogy to the CSP, Chirasil-Val [16]. Thermal immobilization of Chirasil-β-Dex onto a fused silica surface provides universal applicability of this CSP in different chromatographic and electromigration modes. The immobilization of Chirasil-β-Dex on surfaces is a prerequisite for its universal use in all enantioselective chromatographic methods (packed column and open tubular columns), including a unified approach (vide infra) [17].

GC capillary columns coated with derivatized CDs have been used for enantiomeric analysis in many fields, including control of authenticity of essential oils, flavors, and fragrances; clinical and pharmaceutical chemistry; enzymatic reactions; and environmental matrices (organochlorine pesticides, alkyl nitrates, etc.) [18]. For example, 50% heptakis-(6-O-TBDMS-2,3-di-O-methyl)-β-CD dissolved in OV1701 and CP-Chirasil-Dex CB β-CD bound to dimethylpolysiloxane were used for the enantiomeric separation of polychlorinated pesticides. Chlordane components and metabolites have also been analyzed by high resolution GC using a fused capillary column coated with OV1701 containing 35% TBDM as chiral selector, with electron-capture negative ion (ECNI) mass spectrometric detection. Typical chromatograms showed that the CD-coated stationary phases had good separation characteristics [19, 20]. Preparative enantioselective GC has been employed for the chiral separation of xenobiotics on columns coated with 20% TBDM in polysiloxane OV 1701. Atropisomeric polychlorinated biphenyls (PCBs) could be separated on a fused-silica capillary column coated with 25% β-BSCD in a mixture of diphenyl and dimethylpolysiloxane (15:85). The good separating power of the method makes it suitable for the atropisomeric separation of PCBs, even in biological matrices [21]. One new CD derivative, octakis(2,3-di-O-methoxymethyl-6-O-tert-butyldimethylsilyl)-γ-cyclodextrin (2,3-MOM-6-TBDMS-γ-CD), was found to be a suitable chiral stationary phase for enantiodifferentiation of a broad spectrum of chiral volatiles from different chemical classes, most of which are used as flavoring and fragrance materials. A total of 125 pairs of enantiomers were separated. High α values up to 1.8 were observed for the hydroxyketone acetoin and some methyl branched ketones [22].

The database "ChirBase/GC" provides method information for over 24000 enantioseparations of more than 8000 chiral molecules. The database contains experimental conditions, along with details about structures, substructures, and structural similarities [23].

8.3.2 *CDs in liquid chromatography*

α-, β- and γ-CDs are suitable for different applications because of their different cavity diameters. In general, α-CD is not suitable for the selection of larger molecules because of its narrow cavity, and β-CD, which has a larger cavity diameter, has found a wider range of applications in thin-layer chromatography (TLC) separations. Although β-CD is inexpensive, there are limitations in the use of native β-CD as a mobile phase additive because of its low solubility. Attempts to increase the solubility of β-CD have included synthetic modifications and the addition of large amounts of urea to the mobile phase to increase its solubility in water. In recent decades, CDs and CD derivatives have found extensive application in liquid chromatography (LC) to improve the separation of compounds with highly similar chemical structures (mainly positional and optical isomers).

8.3.2.1 *CDs in thin-layer chromatography (TLC)*

Although the separation capacity of TLC, even of high-performance TLC, is generally lower than that of HPLC under similar conditions, TLC remains popular for routine and spot analyses because of its simplicity and low cost. There are many examples of the application of CDs and their derivatives in TLC separations, as components of both mobile and stationary phases.

Aqueous CD solutions have some advantages compared with traditional organic solvents as mobile phases in TLC, including improvement of selectivity and enhancement of chromatographic detection. Hinze *et al.* [24] studied the separation of 25 different phenols and naphthols and 18 substituted benzoic acid derivatives by TLC, using polyamide plates as the solid phase and α-CD in the mobile phase. The *o*-, *m*- and *p*-isomers of nitrophenol, which were easily embedded by α-CD, had larger R_f values and values generally decreased in the order *p*-> *m*-> *o*-substituted isomers. Lepri *et al.* [25] investigated the separation of methylthiohydantoin derivatives of D- and L-amino acids and various naphthyl derivatives on SilC18-50F plates, using β-CD as the chiral agent in an aqueous–organic mobile phase. Optical isomers of dansyl-, dinitrophenyl-, dinitropyridyl- and α-naphthylamide-substituted amino acids were completely separated. It has been reported that isomeric *o*-, *m*- and *p*-substituted benzenes, pesticides, polycyclic aromatic hydrocarbons (PAHs), and drug test mixtures could also be differentiated by TLC on a polyamide stationary phase using an aqueous solution of urea-solubilized β-CD as the mobile phase.

Joseph *et al.* [26] investigated reversed phase (RP) TLC separations of enantiomers of dansyl-amino acids using β-CD as a chiral selector in the mobile phase. After conversion to 5-dimethylamino-1-naphthalene sulfonyl (dansyl) derivatives, the enantiomers of nine racemic proteinogenic amino acids were separated by TLC.

The separation of two types of bile acid, chenodeoxycholic acid (CDCA) and deoxycholic acid (DCA), as well as their conjugated derivatives, by TLC has been a long-standing challenge. Momose *et al.* [27] developed a method for the separation of unconjugated CDCA and DCA and their conjugated derivatives using two-dimensional RP TLC with 5 mM heptakis-(2,6-di-O-methyl)-β-CD (DIMEB). A high degree of separation of individual bile acids in each homologous series was achieved using aqueous methanol in the first dimension and the same solvent system containing DIMEB in the second dimension.

8.3.2.2 *CDs in high performance liquid chromatography (HPLC)*

Because of their versatility, sensitivity and reproducibility, HPLC methods using CD additives in both the stationary and mobile phases have been extensively used for the separation of optical and positional isomers. For example, CDs and their derivatives have been used in the separation of positional and optical isomers of pharmaceuticals, amino acids and related compounds, pesticides and other environmental pollutants, and miscellaneous organic compounds.

8.3.2.2.1 CD-bonded stationary phases

CD-based chiral stationary phases (CSPs) have been successfully used for the resolution of over 1000 racemic compounds, as well as numerous diastereomers and structural isomers. More than 400 papers describing the use of CD-based CSPs in LC have been published since 1983. The development of robust CSPs with facile synthetic methodologies and reproducible application calls for structurally well-defined CD bonded onto supports with robust and solvent-stable covalent linkers.

Depending on the type of covalent linker between CD chiral selector and silica support, CSPs are herein classified as amide-bonded, amine-bonded, ether-bonded, urea-bonded, or triazole-bonded CD-based CSPs.

Chemically bonded CD-based CSPs based on urethane linkage. Fujimura prepared five series of CSPs by bonding native and carbamoylated β-CD or γ-CD onto silica supports using a urethane covalent linkage [28]. Their ability to separate enantiomers was evaluated using racemic dansylated α-amino acids as analytes. The native γ-CD-bonded phase was found to be very effective for the separation of not only the said enantiomers but also homologues of dansylated amino acids. Under RP conditions, both native and derivatized β-CD-bonded stationary phases have been successfully used in HPLC for the separation of a wide variety of positional, structural, and optical isomers [29, 30].

Wada's group [31] described a facile procedure for preparing a bifunctional β-CD-based CSP (PhCD CSP) by immobilizing per-phenylcarbamoylated β-CD onto silica with a single urethane covalent bond. One of the C6-hydroxyl groups of β-CD was used to construct the urethane linkage between CD and the silica support, while the remaining 6 primary and 14 secondary hydroxyl groups on the larger opening of the CD cone were all derivatized to form phenylcarbamoyl groups via reaction with phenylisocyanate. The presence of the phenyl moieties produced a hydrophobic cluster above the larger opening of the CD cone. Formation of this cluster conferred bifunctionality to the derived stationary phase, which was able to form chiral inclusion complexes in the CD cavity and also hydrophobic interactions at the phenyl cluster. The additional hydrophobic interaction between the phenyl cluster and hydrophobic parts of the analytes improved the resolution of phenylalkyl alcohols and amines that had a hydrophobic aromatic ring and a chiral carbon atom several Angströms distant from the ring. For instance, PhCD CSP demonstrated good separation of β-blocker drugs, such as propranolol and its ester derivatives, in aqueous mobile phases.

In order to expand the range of applications of derivatized β-CD phases in enantioseparation, Iida *et al.* prepared one native and four phenylcarbamoylated β-CD bonded-phase HPLC columns and investigated their ability to separate enantiomers of phenylthiocarbamoyl amino acids (PTC-AAs) [32]. The good enantioseparation of 18 pairs of PTC-AAs and PTC-Gly by the 3.3Ph/CD column further demonstrated that derivatization of some of the hydroxyl groups on the β-CD face with phenyl groups is beneficial for enantioseparation of PTC-AAs. The improved separation is thought to arise from widening of the secondary rim of the CD cone, which enhances penetration into, or interaction with, the β-CD cavity.

Chemically bonded CD CSPs based on urea linkages. Ng's group has developed a new synthetic methodology for immobilization of perfunctionalized CDs onto silica gel via urea covalent linkages, with varying chain lengths of the spacer arms [33, 34]. Perfunctionalized CD derivatives containing an azido moiety were used as key intermediates for the preparation of chemically bonded CD CSPs on silica. Three types of CD CSP were prepared: Type I, with a single urea bond at the C6-position; Type II, with multiple urea bonds at the C6-positions; and Type III, with a single urea bond at the C2-position. The ability of the three CD CSPs to separate enantiomers was evaluated using analytes comprising racemic aromatic substituted alcohols, flavonoids, acids, weak acids, antihistamines, amines, amides, amino alcohols, β-blockers, alkaloids, and neutral compounds. These analytes were well separated using CD CSPs of Type I, II, or III under normal — and/or reversed-phase conditions [35].

Chemically bonded CD CSPs based on amino linkages. Lai and Ng developed a CSP using mono (6^A-N-allylamino-6^A-deoxy) perphenylcarbamoylated β-CD as the chiral selector immobilized onto silica gel by hydrosilylation [36]. This CSP provided good enantioseparation of a range of racemic compounds. The same group also described a facile synthesis of mono (6^A-N-allylamino-6^A-deoxy) permethylated β-CD, which is expected to show complementary enantioseparation capability to the perphenylcarbamoyl analogue [37].

Chemically bonded CD CSPs based on ether linkages. CSPs in which the CD is covalently bonded via amido, urethane, and amino linkers have been used extensively in HPLC, but these CSPs are usually unstable towards hydrolysis. Zhou and his group have developed a series of CD CSPs with ether linkages that are suitable for enantioseparation using HPLC. They initially prepared two new CSPs, 6-deoxyisopropylimino-β-CD bonded onto silica gel and heptakis[2,6-o-diamyl-6-deoxyisopropylimino]-β-CD bonded onto silica gel by introducing a rigid imino group to β-CD. The presence of the Schiff base moiety in these two CD CSPs afforded better selectivity for a series of amino acid analytes [38]. They then developed the synthesis of other imino-functionalized β-CD CSPs, taking advantage of the ability of the Schiff base to enhance π–π interactions, hydrogen bonding, and the formation of polar–polar interactions between the CSP and analytes [39, 40]. The imino-functionalized β-CD CSP derivatives explored included mono (6-deoxy-N-1-phenylethylimino)-β-CD, mono[6-deoxy-R-($-$)-N-1-(2-hydroxyl) phenylethylimino]-β-CD, heptakis(2,6-o-diamyl-6-deoxy-phenylimino)-β-CD, heptakis[2,6-o-diamyl-(6-deoxy-R-($-$)-N-1-phenylethylimino)]-β-CD, and heptakis[6-deoxy-(R-($-$)-N-1-(2-hydroxyl)-phenylethylimino)]-β-CD. Their chromatographic properties were evaluated using a wide range of structurally diverse racemic compounds, including disubstituted benzenes, amino acids, aromatic alcohols, and ferrocene derivatives, with the latter effectively showing baseline resolution. All of the CSPs demonstrated outstanding ability to enantiomerically separate ferrocene analytes, with R values in the range 1.07–8.16.

Chemically bonded CD CSPs based on triazolyl linkages. Click chemistry is a versatile approach in organic synthesis. Kacprzak prepared a novel triazolyl-bonded β-CD CSP via click chemistry of an azido-β-CD with alkyne-modified silica [41], and Dai *et al.* [42] used click chemistry to develop β-CD derivative-based CSPs with a triazolyl linkage, using a soluble organic copper (I) catalyst. These novel CSPs exhibited good stability and excellent enantioselectivity. Since single triazolyl linkers are potentially unstable, particularly after long exposure to buffer, CD CSP with multiple covalent triazolyl bonds were then prepared using "multiple click" reactions of heptakis (6-deoxy-6-azido)-β-CD and heptakis (6-deoxy-6-azido-perphenylcarbamoylated)-β-CD with ω-alkynyl-functionalized silica. These "click"-immobilized perphenylcarbamoylated and permethylated CSPs were evaluated under reversed phase conditions in HPLC and demonstrated good enantioseparation of a wide range of racemic analytes, including aryl alcohols, carboxylic acids, dansylamino acids, and flavonoids. Different derivatization on the CD selector afforded quite different enantiorecognition properties due to differences in π–π, dipole–dipole, and hydrophobic interactions with analyte molecules.

8.3.2.2.2 Aqueous CD solution as mobile phase

Although chiral stationary phases based on CDs and CD derivatives have been successfully employed for the enantiomeric separation of a wide variety of chiral compounds, each stationary phase is only suitable for the separation of a limited number of chiral compounds. The use of CDs as mobile-phase additives can partially or entirely overcome this difficulty. Separations can generally be carried out on a traditional reversed phase column by varying the chromatographic conditions, such as the type and concentration of CD, pH, ionic strength, and composition of mobile phase. HPLC methods using CDs as mobile phase additives are thus versatile and easy to carry out [43].

The highly selective chromatographic separations that can be achieved with a CD-containing mobile phase are attributable to the difference in the stability constants of inclusion complexes in the mobile-phase solution. The use of a CD-containing mobile phase has other significant advantages compared with traditional organic solvent systems, including low toxicity and volatility, solubility, and the ability to simultaneously separate both nonpolar and polar solutes.

CD-containing mobile phases have been successfully used in HPLC for the separation of structural isomers, diastereomers, and enantiomers [44]. For instance, CDs have frequently been used as mobile-phase additives for the analysis of pharmaceuticals, such as steroidal drugs, salicylic acid derivatives, antidepressants, tranquilizers, and antitumor agents. Impurities in salicylic acid could be separated on a phenyl column or C18 column using β-CD as a mobile-phase additive [45]. The chromatograms indicated that β-CD in the eluent improved separations on both columns. The enantiomers of terfenadine, a nonsedating histamine H1-receptor antagonist, have also been separated by RP-HPLC, using β-CDs as mobile-phase additives. Sulfobutyl ether-β-CD (SEB-β-CD) was successfully used for chiral separation of the calcium channel blocker rac-amlodipine. It was assumed that formation of inclusion complexes and ion-pairing interactions between the positively charged enantiomers and the negatively charged SBE-β-CD accounts for the resolution [46].

A semi-preparative HPLC technique, using β-CD as a chiral selector in the mobile phase, has been used to separate the enantiomers of the antihistaminic drug brompheniramine. It was reported that the method can also be used for the semipreparative isolation of other pairs of enantiomers [47].

New CD derivatives extend the range of chiral interactions and are able to separate a much wider variety of compounds than native CDs. They are competitively priced and more stable, even allowing reversed phase operation [48]. The preparative capacity of these new mobile-phase additives is also better than that of native CDs in cases of improved chiral recognition. These CD derivatives can be modified by acetylation or S-hydroxypropylation or by formation of the S- or R-naphthyl ethylcarbamate, the 3,5-dimethylphenyl carbamate, or the cyclobond PT para-toluoyl ester.

8.3.3 *CDs in affinity chromatography*

CDs are known to interact with enzymes, especially amylolytic enzymes, such as α-amylases, β-amylases, α-glucosidases, pullulanases, and cyclodextrin glycosyltransferases, and these enzymes can, therefore, be purified by CD affinity chromatography [49, 50].

For amylases that digest raw starch, including α-amylases, amyloglucosidases, and cyclodextrin glycosyltransferases, CDs had no effect on the ability of most of these enzymes to hydrolyze soluble starch but inhibited the digestion of raw starch. This demonstrates that when CD affinity chromatography is used to purify amylases that digest raw starch, it is most likely that the enzyme binds to the CD column through an affinity site. On the other hand, β-amylases and pullulanase-type amylases were competitively inhibited by CDs and, therefore, the use of CD affinity chromatography in the purification of these enzymes is based on interactions between the CDs and the active sites [49].

CDs exhibit different affinities for different amylolitic enzymes [51], and the enzymes can thus be eluted from affinity columns with different concentrations of CDs. For example, α-CD inhibits the activity of β-amylase rather than α-amylase, according to the interaction between immobilized α-CD and the active site on β-amylase. An α-CD Sepharose 6B column was developed to separate β-amylase from α-amylase since β-amylases bound to the column whereas α-amylases did not. The β-amylase purified by this method had a higher activity because of the purification. Similar separation behavior emerges in the case of glycogen debranching enzymes.

As another example, β-CD tetradecasulfate has a very strong affinity for fibroblast growth factor (FGF). Chromatography systems with a β-CD tetradecasulfate stationary phase could, therefore, be used for purification of FGF. It was reported that crude FGF from rat chondrosarcoma could be purified 200000-fold using this method [52].

8.3.4 *CDs in supercritical fluid chromatography*

Supercritical fluid chromatography (SFC) is a separation technique similar to HPLC. The mobile phase is a binary or ternary mixture with CO_2, which is environmentally friendly and has a high diffusion coefficient, as the main component. Because of its considerable advantages, including higher efficiency, higher throughput, more rapid equilibration and shorter cycle times, SFC combined with CDs has found application in the analysis of bioactive compounds and other organic molecules, particularly chiral analytes [53].

CD-oligosiloxane copolymers have been successfully used for the enantiomeric separation of a wide variety of racemic compounds in open tubular column SFC [54]. The use of naphthylethylcarbamoylated-β-CD stationary phases (NEC-CD CSP) in SFC gave good enantiomeric separation, with a shorter analysis time compared with HPLC [55]. Permethyl-substituted β-CD polysiloxane stationary phases have also been synthesized and showed excellent enantioselectivity for a wide variety of racemic compounds on fused silica capillary columns in both GC and SFC separations [56]. The enantiomers of ten racemic benzodiazepines were separated by open tubular SFC, using permethylated β-CD as chiral selector. Good enantiomeric separation of dihydrodiazepam, oxazepam, lorazepam and ethyl loflazepate was achieved using SFC. Octakis-(3-O-butanoyl-2,6-di-O-n-pentyl)-γ-cyclodextrin was covalently bonded to a polysiloxane using an octamethyl spacer (Chirasil-γ-Dex) and showed good enantiomeric selectivity when used as the stationary phase in SFC [57].

In practice, all of the CD chiral stationary phase-based packed columns, which give excellent chiral separation in HPLC, are also applicable to SFC. The high versatility of SFC makes it a valid alternative to both normal and reversed phase HPLC for the enantiomeric separation of racemic compounds.

Another important application of CD derivatives in SFC is their use as additives in chiral mobile phases (CMP) for enantiomeric separations. Salvador and coworkers [58] investigated the enantioseparation of several racemates using dimethyl-β-CD (DM-β-CD) as CMP in SFC. Addition of DM-β-CD to the SFC mobile phase, with acetonitrile or methanol as the polar modifier, provided the highest retentions, efficient chiral separations, and high enantioselectivities.

8.3.5 *CDs in electrophoresis*

Because of high separation efficiencies and very small injection volumes, modern capillary electrophoretic techniques, including capillary electrophoresis (CE), capillary isotachophoresis (CITP), capillary electrochromatography (CEC), micellar electrokinetic chromatography (MEKC), and capillary isoelectric focusing (CIEF), have become well-established analytical methods, not only for the separation of large biomolecules but also for the analysis of small solutes [59].

In 1982, CDs were first used as complexing additives in isotachophoretic analysis of alkali and alkaline metals and markedly improved the separation. Since that time, many investigations into the use of CDs in various types of CE have been carried out [60]. CDs are usually applied as additives to the background electrolyte (BGE) and represent a pseudo phase as they are not fixed in the system. Besides the choice of an appropriate CD, it is also necessary to carefully select the BGE, taking into account factors such as type of buffer and pH, ionic strength and (in some cases) a suitable additional organic solvent to improve the solubility of analytes to achieve the required chiral separation. In the case of water-insoluble analytes, fully nonaqueous CE can provide a solution [61].

Chiral CE separations using CDs are based on two mechanisms: chromatographic and electrophoretic. The ability of CDs to stereoselectively recognize individual enantiomers of a specific analyte is a chromatographic principle, although CDs are not fixed in the separation system in CE. CDs are dissolved in the BGE and thus create a pseudo phase. On the other hand, the analyte, in the free form or complexed with CD, is driven in CE by an electrophoretic principle governed by the charge density of the moving species. As a consequence, the term electrokinetic chromatography (EKC) more accurately describes the real situation than CE [62].

Although most early work exploring applications of CDs in enantiomeric separations by CE centered on the parent α-, β-, and γ-CDs; derivatized neutral CDs quickly gained popularity. The derivatived β-CDs not only offer better enantioselectivity than the parents but also provide much greater solubility in aqueous buffer systems. The most commonly used neutral derivatives of β-CD are the methyl, dimethyl, trimethyl, hydroxyethyl, and hydroxypropyl derivatives [63]. Because of their zero charge, native CDs can only be used for the separation of charged analytes. This problem was partially solved by the application of micellar electrokinetic chromatography (MEKC) and brought one significant advantage since these CDs have only a slight influence on the conductivity of the BGE. Native CDs are often used for the separation of compounds that have an aromatic moiety or a long hydrocarbon chain in their structure [2]. The methyl CD derivatives are effective in separating chiral compounds, such as the enantiomers of terbutaline, ephedrine, and carnitine. The hydroxyalkylated β-CDs are also frequently used as chiral selectors for enantiomeric separations. Neutral randomly substituted 2-hydroxypropylated β-CD has been immobilized on the inner capillary wall and then used for open tubular CEC. A mixture of water-soluble and fat-soluble vitamins or esterom was successfully separated by MEKC by the addition of γ-CD to the SDS electrophoretic medium [64].

In recent years, charged CDs have become increasingly popular for the separation of enantiomers by CE. When employing charged CDs as chiral selectors, the type and concentration of the chiral selector, ionic strength, pH, composition of BGE and organic modifier all have important effects on the efficiency of enantiomeric separation. Altering the charge state of the analyte and chiral selector can affect both the degree and enantioselectivity of inclusion [65].

Anionic CDs are useful for the separation of enantiomers of positively charged compounds because they provide a means of maximizing the electrophoretic mobility difference between the free and complexed forms of the guests. The optimum concentration of the charged CD was much lower while the binding constants were much larger. Carboxymethylated CDs and sulfated CDs are important commercially available randomly substituted negatively charged CDs [66]. The chiral separation of catecholamines (norepinephrine, epinephrine, and DOPA) and their precursors (phenylalanine and tyrosine) was performed using anionic sulfated β-CD (SCD) as chiral selector. The concentration of SCD exerted a decisive influence on the enantiomeric separation. The electropherograms indicated that a higher concentration of SCD shortened migration time and improved resolution [67]. When enantiomeric separations were carried out in a coated capillary at room temperature, with a running buffer consisting of 100 mM KH_2PO_4 (pH 7.5) containing different concentrations of heptakis-(2,3-di-O-acetyl)-β-CD, a series of phenethylamines was separated completely by acetylated β-CDs. Another example of the benefit of using an anionic CD with a cationic analyte is the separation of atenolol enantiomers in a fused silica capillary using a lithium phosphate buffer at pH 2.2. Superior resolution of the atenolol enantiomers was achieved using sulfobutylether-β-CD [68]. Randomly substituted sulfated CDs have been successfully used for both the enantioseparation of 56 types of drug, including anesthetics, antiarrhythmics, antidepressants, antihistamines, and antimalarials, and the chiral separation of the catecholamines, norepinephrine and epinephrine, and their precursors, phenylalanine and tyrosine [69].

Compared with anionic CD derivatives, cationic CD derivatives have advanced more slowly because of their absorption onto column walls and more complicated synthetic procedures. They do, however, demonstrate an advantage over anionic CDs in that they reduce the migration time of solutes in CE. Cationic CDs can be either strong electrolytes, in which case they are functionalized with quaternary ammonium groups, or weak electrolytes [70]. Cationic CDs contain alkyl sulfonate or other acidic groups and are useful for the separation of enantiomers of positively charged compounds because they provide a means of maximizing the electrophoretic mobility difference between the free and complexed forms of the guests [71]. The only commercially available positively charged randomly substituted CD is quaternary ammonium β-CD (QA-β-CD), which has been used for the separation of negatively charged analytes. A highly water-soluble CD derivative, 6-O-(2-hydroxy-3-trimethylammonio-propyl)-β-CD (6-HPTMA-β-CD), was synthesized and successfully used as a chiral selector for the enantiomeric separation of some acidic compounds using capillary zone electrophoresis (CZE) in an uncoated capillary [72]. Because HPTMA-β-CD is as basic as QA-β-CD, it can also adsorb onto the walls of the capillary under the general CE experimental conditions used for enantioseparation [73].

8.4 Applications of CDs in spectroscopy

The high electron density inside the CD cavity can grip the electrons of encapsulated guest molecules and produce a significant and temporary change of the analyte microenvironment (from aqueous hydrophilic to lipophilic). This results in an enhanced spectral response of both the guest and host [74]. The CD cavity is a unique microenvironment, similar to that of organic solvents, which can alter parameters such as spectral shifts and excited state lifetimes. The effects of CDs on the spectral properties of guest molecules have led to their utilization in various spectrometric techniques, including UV/VIS spectrophotometry, fluorescence and luminescence spectroscopy, and NMR spectroscopy.

8.4.1 *CDs in UV/VIS spectrophotometry*

CDs can change the UV/VIS spectra of guest molecules by complexing the guest. Usually, the intensity and position of the absorption bands in the spectrum are altered after formation of an inclusion complex. Because of the hydrophobic cavity of CDs, the spectra of included guests in aqueous solutions are very similar to those of guests in organic solvents. In UV/VIS spectrophotometry, CDs are used mainly to improve the solubility and stability of colored compounds or to increase the sensitivity and selectivity of color reactions. CDs could, therefore, be used as reagents to improve the sensitivity of determinations in UV/VIS spectrometry and could be useful auxiliaries in the spectrophotometric determinations of a wide variety of compounds and elements [75]. The remarkable spectral changes in UV/VIS spectra upon adding CDs have also been used to calculate dissociation constants with the Scott equation or the Benesi–Hildebrand equations.

β-CD has been reported to improve the selectivity of the color reactions of various metal ions with triphenylmethane, xanthene acid dyes, and some other coloring reagents. The effect of β-CD on the association compound system of metal (Mo, Zn, Co)-thiocyanate basic dyes, such as malachite green, crystal violet, rhodamine B, rhodamine 6G, and butyl rhodamine B, has been investigated. β-CD was found to contribute to a more sensitive and stable system, which improved the solubility of the basic dyes and produced a favorable microenvironment for the color reactions [76]. β-CD has also been used to solubilize 1,2-diaminoanthraquinone in water. The inclusion complex that is formed acts as a ligand for metal ions and could be used for the determination of trace levels of palladium by spectrophotometry. In the spectrophotometric determination of microamounts of Zn based on the Zn-dithizone color reaction, β-CD increased the apparent molar absorptivity at 538 nm by 8.37-fold. In the spectrophotometric analysis of copper in leaves based on the color reaction of Cu(II) and *meso*-tetrakis (4-methoxy-3-sulfophenyl) porphyrin, the presence of α-CD enhanced the sensitivity by 50% [77, 78]. CDs also can be used to increase the stability of color indicators used for the spectrophotometric determination of hydrogen peroxide in body fluids. CDs and their derivatives also have applications in enzyme assays and measurement of enzyme activity. For example, glucosyl- and maltosyl-α-CD have both been used to increase the accuracy and sensitivity of amylase assays.

8.4.2 *CDs in luminescence spectroscopy*

Molecular luminescence spectrometry, especially molecular fluorescence spectrometry, has become established as a routine technique in many analytical applications. In many cases, molecular luminescence spectrometry provides a lower detection limit and greater selectivity than molecular absorption spectrometry. Most organic compounds, however, show enhanced fluorescence or phosphorescence in ordered organic systems, e.g., in lipophilic solvents, whereas the intensity of luminescence of the same compounds can be poor in hydrophilic environments, e.g., in water. In aqueous solution, the relatively lipophilic nanocavities of CDs offer a local microenvironment that results in useful enhancement of the fluorescence or phosphorescence signals of complexed fluorescent analytes. The CD cavity behaves like an organic solvent, providing apolar surroundings for the entrapped chromophores. This altered lipophilic microenvironment can provide favored polarity and acid–base equilibria, leading to enhanced quantum efficiencies and increased luminescence intensity.

The polarity of the actual microenvironment inside the CD cavity is similar to that of some oxygen-containing organic solvents such as 1,4-dioxane, *tert*-amyl alcohol or 1-octanol. The CD-complexed analyte thus experiences significantly lower polarity in the local cavity environment. Reduced freedom of movement in the cavity also results in hindered intramolecular rotations [79]. CDs also protect the fluorescing singlet state or the phosphorescing triplet state of the analyte from external quenchers, and the rotation of the guest molecule becomes hindered. Both of these effects may result in a decrease in vibrational deactivation [80, 81].

The fluorescence intensities of many compounds, including pyrene, various illicit drugs, narcotics, hallucinogens, and polychlorinated biphenols, are significantly increased by formation of complexes with CDs and their derivatives. The molecular probe 1-anilinonaphthalene-8-sulfonate, which shows very weak fluorescence in water, can exhibit tenfold stronger fluorescence after adding CDs to the aqueous solution. The addition of CDs to aqueous solutions of compounds such as ammonium 7-fluorobenzo-2-oxa-1,3-diazole-4-sulfonate-labeled glutathione, acetylcysteine, and some dansylated amino acids, increased the fluorescence emission about eightfold, compared with the original values [82]. The fluorescence intensity of naphthalene in aqueous solution decreases upon aeration but the quenching is totally suppressed in the presence of a water-soluble sulfopropylated β-CD [83]. Similarly, the quenching of halonaphthalene phosphorescence in water by $NaNO_2$ can be substantially inhibited by β-CD. The rate of inhibition depends on the tightness of the bonds between the analyte and the CD. Retinal, which is normally insoluble in water and is not fluorescent in solution at room temperature, emits luminescence in the region of 450 nm and permits fluorescence detection when incorporated into β- or γ-CD, even in air-saturated aqueous solution [84]. CDs can be used as solid matrices for the detection of volatile compounds by luminescence. The CDs trap the volatile compounds and allow measurement of room-temperature fluorescence and room-temperature phosphorescence from the absorbed compounds [85]. This approach gave subpicogram detection limits and well-resolved spectra in the determination of polynuclear aromatic hydrocarbons, nitrogen-containing heterocycles, and bridged biphenyls.

CDs have also been used to enhance chemiluminescence of luminol-related compounds. CDs increased the light output up to 300-fold in aqueous peroxyoxalate solution. The enhancement was attributed to increases in reaction rate, excitation efficiency, and fluorescence efficiency of the emitting species. In most cases, the presence of CDs will enhance luminescence, but CDs can also selectively quench the luminescence of some compounds. A study of the effect of β-CD on the fluorescence of xanthene dyes, coumarins, and pyromethened-ifluoroboron complexes in aqueous solution showed that β-CD enhanced the fluorescence of 7-hydroxy-coumarin and coumarins but quenched the fluorescence of 7-hydroxy-4-methyl-coumarins [86]. This behavior of CDs provides a new approach for multicomponent fluorometric analysis and allows CDs to be used for differentiating the structures of similar compounds, such as positional isomers, by the selective incorporation of the analyte. Recently, supramolecular systems of sensitizers *meso*-tetrakis(4-sulfonatophenyl)porphine (TPPS4), its zinc(II) and palladium(II) complexes and *meso*-tetrakis(4-carboxyphenyl) porphine with native CD and HP β-CD have been studied. The results indicate that there are strong noncovalent host–guest interactions between selected sensitizers and CDs (association constants, $K \sim 10^7 \, M^{-1}$) and that complexation significantly effects the physical and photophysical properties of these compounds [87].

Besides synthetic food contaminants, plant toxins, food additives, and agrochemical residues, mycotoxins are considered to be one of the most significant chronic dietary risk factors. Analysis of mycotoxins has traditionally been performed using mainly chromatographic techniques, which provide high sensitivity and accuracy. Faster screening assays capable of simultaneous *in situ* multitoxin determination are, however, gaining popularity. One of these newer assays depends on the significant enhancement of mycotoxin fluorescence in the presence of CDs [88]. A portable fluorometer, suitable for rapid and highly sensitive screening of aflatoxins in real-world food samples, was constructed using CD-enhanced fluorescence detection. The compact and easy-to-handle device enabled simple and rapid monitoring of aflatoxins in milk, with an enhanced detection limit. The CD-enabled aflatoxin sensor was suitable for preliminary screening of noncontaminated and contaminated farm milk products at early stages of the production chain and was able to detect analytes at concentrations as low as 25 ppm [89].

8.4.3 *CDs in NMR spectroscopy*

In NMR spectroscopy, CDs are used mainly as chiral NMR shift reagents. Complexation with CDs can alter the characteristics of the NMR spectra of guest molecules. In many cases, the formation of CD inclusion complexes results in separation of the chemical shifts of the two enantiomers in racemic mixtures [90, 91]. ^{1}H NMR, ^{13}C NMR, and ^{15}N NMR spectroscopy have all been used to study CD inclusion complexes and their properties. Such studies have examined the structures of the inclusion compounds; the interaction of CDs with acids, aliphatic amines, and cyclic alcohols; and the optical purity of guests. Both 1D and 2D ^{1}H NMR spectroscopy have been used to investigate the chiral recognition process in CE. The changes in the shifts of the 3H and 5H protons in the ^{1}H NMR spectrum demonstrated the formation of an inclusion complex. When the guest molecule was enclosed in the CD cavity, the resonance signals of protons located inside the cavity (3H and 5H) were shifted in the spectrum. The signals of protons located on the outer side of the cavity (2H, 4H, and 6H) remain relatively unaffected. NMR spectroscopy, from initial ^{1}H NMR solution spectra to ^{13}C NMR, ^{15}N NMR, ^{19}F NMR, and ^{31}P NMR solid-state spectra, has become the most powerful approach for the study of inclusion complexes formed between CDs and a variety of guest molecules. A ^{19}F NMR study on the formation of diastereoisomeric inclusion complexes between fluorinated amino acid derivatives and α-CD in 10% D_2O solution showed that the chemical shifts of the D-amino acid derivatives included by α-CD were upfield from those of their L-analogues [92]. The shift difference between the diastereoisomers formed with D- and L-enantiomers can be used for chiral analysis and optical purity determinations. For example, the interaction of β-CD with propanolol hydrochloride produced diastereomeric pairs. The protons of the antipodes produced ^{1}H NMR signals that differed in chemical shifts in D$_2$O solution at 400 MHz. The intensity of the resonance signals for each diastereoisomer has been used for optical purity determination. By adding racemate into pure ($-$) isomer, this technique was able to measure the optical purity of propanolol hydrochloride in water down to the level of 1%.

References

1. Szejtli J.(1998). *Cyclodextrin Technology*. Boston: Kluwer Academic Publishers.
2. Szejtli J. (1998). Introduction and general overview of cyclodextrin chemistry. *Chemical Reviews*, 98, p. 1743.
3. Liu L, Song KS, Li XS. (2001). Charge-transfer interaction: a driving force for cyclodextrin. Inclusion complexation. *Journal of Inclusion Phenomena and Macrocyclic Chemistry*, 40, pp. 35–39.
4. Szejtli J. (2004). Past, present, and future of cyclodextrin research. *Pure Applied Chemistry*, 76(10), pp. 1825–1845.
5. Li S, Purdy WC. (1992). Cyclodextrins and their applications in analytical chemistry. *Chemical Reviews*, 92, pp. 1457–1470.
6. Mosinger J, Viktorie T, Irena N, *et al.* (2001). Cyclodextrins in analytical chemistry. *Analytical Letters*, 34(12), pp. 1979–2004.
7. Moon JY, Jung HJ, Moon MH. (2008). Inclusion complex-based solid-phase extraction of steroidal compounds with entrapped β-cyclodextrin polymer. *Steroids*, 73, pp. 1090–1097.
8. Liu H, Liu C, Yang X, *et al.* (2008). Uniformly sized β-cyclodextrin molecularly imprinted microspheres prepared by a novel surface imprinting technique for ursolic acid. *Analytica Chimica Acta*, 628, pp. 87–94.
9. Lantz AW, Rodriguez MA, Wetterer SM, Armstrong DW. (2006). Estimation of association constants between oral malodor components and various native and derivatized cyclodextrins. *Analytica Chimica Acta*, 557, pp. 184–190.
10. Morgan BR, Artiss JD, Zak B. (1998). Determination of total cholesterol in hypertriglyceridemic serums. *Microchemical Journal*, 59(2), 315–322.
11. Juvancz Z, Szejtli J. (2002). The role of cyclodextrins in chiral selective chromatography. *TrAC Trends in Analytical Chemistry*, 21(5), 379–388.
12. Berthod A, Li W, Armstrong DW. (1992). Multiple enantioselective retention mechanisms on derivatized cyclodextrin gas chromatographic chiral stationary phases. *Analytical Chemistry*, 64(8), pp. 873–879.
13. Armstrong DW, Li W, Stalcup AM, *et al.* Capillary gas chromatographic separation of enantiomers with stable dipentyl-α, β- and γ-cyclodextrin-derivatized stationary phases. *Analytica Chimica Acta*, 234(2), pp. 365–380.
14. Blum W, Aichholz R. (1990). Gas chromatographic enantiomer separation on tert-butylsilylated β-cyclodextrin diluted in PS-086. A simple method to prepare enantioselective glass capillary columns. *Journal of High Resolution Chromatography*, 13, pp. 515–518.

15. Schurig V. (2002). Chiral separations using gas chromatography. *TrAC Trends in Analytical Chemistry*, 21, pp. 647–661.
16. Frank H, Nicholson GJ, Bayer E. (1977). Rapid gas chromatographic separation of amino acid enantiomers with a novel chiral stationary phase. *Journal of Chromatographic Science*, 15, pp. 174–176.
17. Schurig V, Schmalzing D, Muhleck U, *et al.* (1990). Gas chromatographic enantiomer separation on polysiloxane-anchored ermethyl-β-cyclodextrin (Chirasil-Dex). *Journal of High Resolution Chromatography*, 13, pp. 713–717.
18. Schurig V. (2001). Separation of enantiomers by gas chromatography. *Journal of Chromatography A*, 906, pp. 275–299.
19. Kim BE, Lee KP, Park KS. Enantioselectivity of 6-O-alkyldimethylsilyl-2,3-di-O-methyl-β-cyclodextrins as chiral stationary phases in capillary GC. *Chromatographia*, 46(3–4), pp. 145–150.
20. Klobes U, Vetter W, Luckas B, *et al.* (1998). Enantioseparation of the compounds of technical toxaphene (CTTs) on 35% heptakis (6-O-tert-butyldimethylsilyl-2,3-di-O-Methyl)-β-cyclodextrin in OV1701. *Chromatographia*, 47(9–10), pp. 565–569.
21. Vetter W, Klobes U, Luckas B, *et al.* (1997). Enantiomer separation of selected atropisomeric polychlorinated biphenyls including PCB 144 on tert-butyldimethylsilylated β-cyclodextrin. *Journal of Chromatography A*, 769(2), pp. 247–252.
22. Takahisa E, Karl-Heinz E. (2005). 2,3-Di-O-methoxymethyl-6-O-tert-butyl dimethylsilyl-γ-cyclodextrin: a new class of cyclodextrin derivatives for gas chromatographic separation of enantiomers. *Journal of Chromatography A*, 1063(1–2), pp. 181–192.
23. http://www.acdlabs.com/products/chrom lab/chirbase/gc.html.
24. Hinze WL, Pharr DY, Fu ZS, *et al.* (1989). Thin-layer chromatography with urea solubilized .beta.-cyclodextrin mobile phases. *Analytical Chemistry*, 61(5), pp. 422–428.
25. Lepri L, Boddi L, Del Bubba M, Cincinelli A. (2001). Reversed-phase planar chromatography of some enantiomeric amino acids and oxazolidinones. *Biomedical Chromatography* (Special Issue: Chiral Resolution), 15(3), pp. 196–201.
26. Günther K, Richter P, Möller K. (2003). Separation of enantiomers by thin-layer chromatography: an overview. In: Güitz G, Schmid MG (Eds.), *Chiral Separations Methods and Protocols*, Vol. 243, Methods in Molecular Biology. Heidelberg: Springer, pp. 29–59.
27. Momose T, Mure M, Iida T, Goto J. (1998). Method for the separation of the unconjugates and conjugates of chenodeoxycholic acid and deoxycholic acid by two-dimensional reversed-phase thin-layer chromatography with methyl betacyclodextrin. *Journal of Chromatography A*, 811(1–2), pp. 171–180.
28. Fujimura K, Suzuki S, Hayashi K. (1990). Retention behavior and chiral recognition mechanism of several cyclodextrin-bonded stationary phases for dansyl amino acids. *Analytical Chemistry*, 62, 2198–2205.

29. Guillaume YC, Robert JF, Guinchard C. (2001). Role of the water as a surface tension modifier for the retention of a series of dansylaminoacids on a β-CD stationary phase. *Talanta*, 55, pp. 263–269.

30. Kim H, Radwanski E, Lovey R. (2002). Pharmacokinetics of the active antifungal enantiomer, SCH 42427 (RR), and evaluation of its chiral inversion in animals following its oral administration and the oral administration of its racemate genaconazole (RR/SS). *Chirality*, 14, pp. 436–441.

31. Nakamura K, Fujima H, Kitagawa H. (1995). Preparation and chromatographic characteristics of a chiral-recognizing perphenylated cyclodextrin column. *Journal of Chromatography A*, 694, pp. 111–118.

32. Iida T, Matsunaga H, Fukushima T, *et al.* (1997). Complete enantiomeric separation of phenylthiocarbamoylated amino acids on a tandem column of reversed and chiral stationary phases. *Analytical Chemistry*, 69, pp. 4463–4468.

33. Zhang LF, Chen L, Lee TC. (1999). A facile route into 6A-mono-ω-alkenylcarbamido-6 A -deoxy-perfunctionalised cyclodextrin: key intermediate for further reactive functionalisations. *Tetrahedron Asymmetry*, 10, pp. 4107–4113.

34. Ng SC, Chen L, Zhang LF, *et al.* (2002). Facile preparative HPLC enantioseparation of racemic drugs using chiral stationary phases based on mono-6A-azido-6A-deoxy-perphenyl carbamoylated β-cyclodextrin immobilized on silica gel. *Tetrahedron Letters*, 43, pp. 677–681.

35. Poon YF. (2006). Chiral enantio-separation of racemic drugs. PhD thesis, National University of Singapore, Singapore.

36. Lai XH, Ng SC. (2004). Preparation and chiral recognition of a novel chiral stationary phase for high-performance liquid chromatography, based on mono (6A-N-allylamino-6A -deoxy)-perfunctionalized β-cyclodextrin and covalently bonded silica gel. *Journal of Chromatography A*, 1031, pp. 135–142.

37. Lai XH, Ng SC. (2004). Convenient synthesis of mono (6A-N-allylamino-6A-deoxy) permethylated β-cyclodextrin: a promising chiral selector for an HPLC chiral stationary phase. *Tetrahedron Letters*, 45, pp. 4469–4472.

38. Zhou ZM, Fang M, Yu CX. (2005). Synthesis and chromatographic properties of α-Schiff bases (6-imino)-β-cyclodextrin bonded silica for stationary phase of liquid chromatography. *Analytica Chimica Acta*, 539, pp. 23–29.

39. Zhou ZM, Li X, Chen XP, Fang M, Dong X. (2010). Separation performance and recognition mechanism of mono(6-deoxy-imino)-β-cyclodextrins chiral stationary phases in high-performance liquid chromatography. *Talanta*, 82, pp. 775–784.

40. Chen XP, Zhou ZM, Yuan H, *et al.* (2008). Preparation and chiral recognition of a mono [6A-N-1-(2-hydroxy)-phenylethylimino-6A-deoxy]-β-cyclodextrin HPLC stationary phase. *Journal of Chromatographic Science*, 46, pp. 777–782.

41. Zhang YP, Guo ZM, Ye JX, *et al.* (2008). Preparation of novel β-cyclodextrin chiral stationary phase based on click chemistry. *Journal of Chromatography A*, 1191, pp. 188–192.

42. Wang Y, Xiao Y, Tan TTY, *et al.* (2008). Click chemistry for facile immobilization of cyclodextrin derivatives onto silica as chiral stationary phases. *Tetrahedron Letters*, 49, pp. 5190–5191.

43. Chen D, Jiang S, Chen Y, Hu Y. (2004). HPLC determination of sertraline in bulk drug, tablets and capsules using hydroxypropyl-β-cyclodextrin as mobile phase additive. *Journal of Pharmaceutical and Biomedical Analysis*, 34, pp. 239–245.
44. Cserháti T, Forgács E. (2003). *Cyclodextrins in Chromatography*. Cambridge: The Royal Society of Chemistry.
45. Goss JD. (1998). Improved liquid chromatography of salicylic acid and some related compounds on a phenyl column. *Journal of Chromatography A*, 828, pp. 267–271.
46. Ovens PK, Fell AF, Coleman MW, *et al.* (1996). Method development in liquid chromatography with a charged cyclodextrin additive for chiral resolution of rac-amlodipine utilising a central composite design. *Chirality*, 8, pp. 466–476.
47. Cooper AD, Jefferies TM. (1993). Semi-preparative high-performance liquid chromatographic resolution of brompheniramine enantiomers using β-cyclodextrin in the mobile phase. *Journal of Chromatography*, 637, pp. 137–143.
48. Xu J, Zhao W, Ning Y, *et al.* (2013). Enantiomer separation of phenyllactic acid by HPLC with Hp-b-cyclodextrin as chiral mobile phase additive. *Journal of Inclusion Phenomena and Macrocyclic Chemistry*, 76, pp. 461–465.
49. Hamilton LM, Kelly CT, Fogarty WM. (2000). Review: Cyclodextrins and their interaction with amylolytic enzymes. *Enzyme and Microbial Technology*, 26(8), pp. 561–567.
50. Kriegshäuser G, Liebl W. (2000). Pullulanase from the hyperthermophilic bacterium *Thermotoga maritima*: purification by β-cyclodextrin affinity chromatography. *Journal of Chromatography B: Biomedical Science and Applications*, 737(1–2), pp. 245–251.
51. Wisessing A, Engkagul A, Wongpiyasatid A, Chuwongkomon K. (2008). Purification and characterization of C. *maculatus* α-amylase. *Kasetsart Journal: Natural Science*, 42, pp. 240–244.
52. Weisz PB, Yardley PA. (1993). Cyclodextrin polymers and cyclodextrins immobilized on a solid surface. US Patent, US5658894 A.
53. Yi G, Bradshaw JS, Rossiter BE, *et al.* (1993). Novel cyclodextrin-oligosiloxane copolymers for use as stationary phases to separate enantiomers in open tubular column supercritical fluid chromatography. *Journal of Organic Chemistry*, 58, pp. 2561–2565.
54. Williams KL, Sander LC, Wise SA. (1996). Comparison of liquid and supercritical fluid chromatography using naphthylethylcarbamoylated-β-cyclodextrin chiral stationary phases. *Journal of Chromatography A*, 746, pp. 91–101.
55. Yi G, Bradshaw JS, Rossiter BE, *et al.* (1993). New permethyl-substituted β-cyclodextrin polysiloxanes for use as chiral stationary phases in open tubular column chromatography. *Journal of Organic Chemistry*, 58, pp. 4844–4850.
56. Mangelings D, Heyden YV. (2008). Chiral separations in sub- and supercritical fluid chromatography. *Journal of Separation Science*, 31, pp. 1252–1273.
57. Williams KL, Sander LC, Wise SA. (1996). Use of a naphthylethylcarbamoylated-β-cyclodextrin chiral stationary phase for the separation of drug enantiomers and related compounds by sub and supercritical fluid chromatography. *Chirality*, 8, pp. 325–331.

58. Salvador A, Herbreteau B, Dreux M. (2001). Preliminary studies of supercritical-fluid chromatography on porous graphitic carbon with methylated cyclodextrin as chiral selector. *Chromatographia*, 53, pp. 207–209.
59. Fanali S. (1996). Identification of chiral drug isomers by capillary electrophoresis. *Journal of Chromatography A*, 735, pp. 77–121.
60. Chankvetadze B. (2007). Enantioseparations by using capillary electrophoretic techniques: the story of 20 and a few more years. *Journal of Chromatography A*, 1168, pp. 45–70.
61. Řezanka, Navrátilová, *et al.* (2014). Application of cyclodextrins in chiral capillary electrophoresis. *Electrophoresis*, 35, pp. 2701–2721.
62. Wang S, Bian L, Tang W. (2013). Cationic cyclodextrins for capillary electrophoresis. In: Tang W, Ng S-C, Sun D (Eds.), *Modified Cyclodextrins for Chiral Separation*. Heidelberg: Springer.
63. Wren S, Berger TA, Boos K-S, Engelhardt H, Adlard ER, Davies IW, Altrai KD, Stock R. (2001). The use of cyclodextrins as chiral selectors. In: *The Separation of Enantiomers by Capillary Electrophoresis*, CHROM Vol. 6, Chromatographia CE Series, Supplement 1. Heidelberg: Springer, pp. s59–s77.
64. Zhou J, Tang J, Tang W. (2015). Recent development of cationic cyclodextrins for chiral separation. *Trends in Analytical Chemistry*, 65, pp. 22–29.
65. Ong CP, Ng CL, Lee HK, Li SFY. (1991). Separation of water- and fat-soluble vitamins by micellar electrokinetic chromatography. *Journal of Chromatography A*, 547, pp. 419–428.
66. Song L, Guo Z, Chen Y. (2012). Separation and determination of chiral composition in penicillamine tablets by capillary electrophoresis in a broad pH range. *Electrophoresis*, 33, pp. 2056–2063.
67. Fakhari AR, Tabani H, Nojavan S, *et al.* (2012). Electromembrane extraction combined with cyclodextrin-modified capillary electrophoresis for the quantification of trimipramine enantiomers. *Electrophoresis*, 33, 506–515.
68. Maruszak W, Schmid MG, Gübitz G, Ekiert E. (2004). Marek trojanowicz separation of enantiomers by capillary electrophoresis using cyclodextrins. *Methods in Molecular Biology*, 243, pp. 275–289.
69. De Boer T, De Zeeuw RA, De Jong GJ. (2000). Recent innovations in the use of charged cyclodextrins in capillary electrophoresis for chiral separations in pharmaceutical analysis. *Electrophoresis*, 21(15), 3220–3239.
70. Boer TD, Zeeuw RA, Ensing K. (2000). Recent innovations in the use of charged cyclodextrins in capillary electrophoresis for chiral separations in pharmaceutical analysis. *Electrophoresis*, 21, pp. 3220–3239.
71. Haynes JL, Shamsi SA, O'Keefe F, *et al.* (1998). Cationic β-cyclodextrin derivative for chiral separations. *Journal of Chromatography A*, 803, pp. 261–271.
72. Lin X, Zhao M, Qi X, *et al.* (2006). Capillary zone electrophoretic chiral discrimination using 6-O-(2-hydroxy-3-trimethylammoniopropyl)-β-cyclodextrin as a chiral selector. *Electrophoresis*, 27, pp. 872–879.
73. Cucinotta V, Contino A, Giuffrida A. (2010). Application of charged single isomer derivatives of cyclodextrins in capillary electrophoresis for chiral analysis. *Journal of Chromatography A*, 1217(7), pp. 953–967.

74. Yu L. (2004). Spectroscopic studies on molecular recognition of modified cyclodextrins. *Current Organic Chemistry*, 8, p. 3546.
75. Wehry EL. (1997). Molecular fluorescence and phosphorescence spectrometry. In: Settle FA (Ed.), *Handbook of Instrumental Techniques for Analytical Chemistry*. Upper Saddle River: Prentice Hall, pp. 507–539.
76. Szente L, Szejtli J. (1998). Non-chromatographic analytical uses of cyclodextrins. *Analyst*, 123, pp. 735–741.
77. León-Rodríguez LMD, Basuil-Tobias DA. (2005). Testing the possibility of using UV–vis spectrophotometric techniques to determine non-absorbing analytes by inclusion complex competition in cyclodextrins. *Analytica Chimica Acta*, 543(1–2), pp. 282–290.
78. Tang B, Ma L, Wang HY, Zhang GY. (2002). Study on the supramolecular interaction of curcumin and β-cyclodextrin by spectrophotometry and its analytical application. *Journal of Agriculture Food Chemistry*, 50(6), pp. 1355–1361.
79. Wagner BD. (2006). The effects of cyclodextrins on guest fluorescence. In: Douhal A (Ed.), *Cyclodextrin Photochemistry, Photophysics and Photobiology*. Oxford: Elsevier, pp. 27–59.
80. Eastwood D. (1985). Advances in luminescence spectroscopy: A symposium. ASTM Committee E-13 on Molecular Spectroscopy.
81. Aicart E, Junquera E. (2003). Complex formation between purine derivatives and cyclodextrins: a fluorescence spectroscopy study. *Journal of Inclusion Phenomena and Macrocyclic Chemistry*, 47(3–4), pp. 161–165.
82. Wagner BD. (1998). The fluorescence enhancement of 1-anilinonaphthalene-8-sulfonate (ANS) by modified β-cyclodextrins. *Journal of Photochemistry and Photobiology A: Chemistry*, 114(2), pp. 151–157.
83. Kano K, Hashimoto S, Imai A, Ogawa T. (1984). Three-component complexes of cyclodextrins. Exciplex formation in cyclodextrin cavity. *Journal of Inclusion Phenomena and Macrocyclic Chemistry*, 2(3–4), pp. 737–746.
84. Maragos CM, Appell M, Lippolis V, Visconti A, Catucci L, Pascale M. (2008). Use of cyclodextrins as modifiers of fluorescence in the detection of mycotoxins. *Food Additives and Contaminants A*, 25(2), pp. 164–171.
85. Kathuria A, Gupta A, Priya N, *et al.* (2009). Specificities of calreticulin transacetylase to acetoxy derivatives of 3-alkyl-4-methylcoumarins: effect on the activation of nitric oxide synthase. *Bioorganic & Medicinal Chemistry*, 17(4), pp. 1550–1556.
86. Dondon R, Fery-Forgues S. (2001). Inclusion complex of fluorescent 4-hydroxycoumarin derivatives with native β-cyclodextrin: enhanced stabilization induced by the appended substituent. *Journal of Physical Chemistry B*, 105(43), pp. 10715–10722.
87. Deumié JM, Kubát KLP, Wagnerová DM. (2000). Supramolecular sensitizer: complexation of meso-tetrakis (4-sulfonatophenyl) porphyrin with 2-hydroxypropyl cyclodextrins. *Journal of Photochemistry and Photobiology A: Chemistry*, 30(1), pp. 13–20.
88. Cucci C, Mignani AG, Dall'Asta C, *et al.* (2007). A portable fluorometer for the rapid screening of M1 aflatoxin. *Sensors and Actuators B: Chemical*, 126, pp. 467–472.

89. Cozzini P, Ingletto G, Singh R, *et al.* (2008). Mycotoxin detection plays "cops and robbers": cyclodextrin chemosensors as specialized police? *International Journal of Molecular Sciences*, 9(12), pp. 2474–2494.
90. Lee S, Yi D, Jung S. (2004). NMR spectroscopic analysis on the chiral recognition of noradrenaline by β-cyclodextrin (β-CD) and carboxymethyl-β-cyclodextrin (CM-β-CD). *Bulletin of the Korean Chemical Society*, 25(2), pp. 216–220.
91. Thomas JW, James DW. (2003). Chiral reagents for the determination of enantiomeric excess and absolute configuration using NMR spectroscopy. *Chirality*, 15(3), pp. 256–270.
92. Brown SE, Coates JH, Lincoln SF, Coghlan DR, Easton CJ. (1991). Chiral molecular recognition: a ^{19}F nuclear magnetic resonance study of the diastereoisomer inclusion complexes formed between fluorinated amino acid derivatives and α-cyclodextrin in aqueous solution. *Journal of the Chemical Society, Faraday Transactions*, 87, pp. 2699–2703.

9. Cyclodextrin-Based Enzyme Mimics

Aiquan Jiao

*State Key Lab of Food Science and Technology,
School of Food Science and Technology,
Jiangnan University,
Wuxi 214122, China*

9.1 Introduction

Enzymes are proteins with high substrate specificities and large rate accelerations, as evolved natural products by several billion years. Enzymes catalyze almost all chemical reactions with stereoselectivities and specificities under mild conditions. The study of natural enzyme aroused the concern of both chemists and biologists, because of its high selectivity, catalytic activity, and mild reactive condition in chemical reaction. Various simple or complicated chemical systems have been widely used as mimic models for natural enzyme [1–5]. Through the investigation of enzyme mimic, the mechanism of the enzymatic catalytic reaction and correlation between the structure, character, and function of the enzyme were explored which should lay a theoretic foundation for designing and synthesizing the catalysts with simple structure, high selectivity, high catalytic activity in mild reactive condition.

A model can represent general features for more than one enzyme! Viewed from a different angle, the requirements necessary for the design of a good enzyme model can be summarized in these five criteria [6]:

(1) Because noncovalent interactions are the key to biological flexibility and specificity, the model should provide a good (hydrophobic) binding site for the substrate.
(2) The model should provide the possibility of forming electrostatic and hydrogen bonds to help the substrate bind in the proper way.
(3) Carefully selected catalytic groups have to be properly attached to the model to effect the reaction.
(4) The structure of the model should be rigid and well-defined, particularly with respect to substrate orientation and stereochemistry.
(5) Of course, the model should preferably be water-soluble, and catalytically active under physiological conditions of pH and temperature.

Cyclodextrins (CDs) are extremely attractive components of artificial enzymes and other biomimetic materials [2, 7]. They are readily available, they bind hydrophobic substrates into their cavities in water solution, and they have two rims of hydroxyl groups that can either react with substrates themselves or be used to attach other catalytic and functional groups. Generally, CD binds to a typical substrate in water with binding constants of 10^2–10^4 M^{-1}. In the case of CD dimer, a substrate can have a binding constant exceeding 10^{11} M^{-1} and is comparable to the binding constants of very strong antibodies [8]. For these reasons, CDs have extensively been exploited as enzyme models and molecular receptors [1, 3, 9, 10]. In this part, we will introduce some typical models in building enzyme mimics on the basis of CDs.

9.2 Structure and catalytic mechanism of biological enzymes

9.2.1 *Structure of biological enzymes*

Enzymes are biocatalysts with a high efficiency and specificity. All biochemical reactions in organism are enzymatic reactions. The high efficiency and selectivity of enzymes originates from their hydrophobic bonding with substrates and the proximity effects of catalytic groups.

Enzyme molecules are composed of the 20 amino acid types, both within the primary and advanced structures. The primary structure refers to the covalent framework of polypeptide chains with certain amino acid sequences. The secondary structure is the helical, folded, angled, and curly microstructures formed by the interaction of the adjacent amino acid residues in the primary structure via hydrogen bonds. The tertiary structure is the spatial arrangement formed by the further distortion based on the secondary structure. The quaternary structure is the spatial distribution with specificity of various subunits. The enzymes with activity are generally globulins. Polypeptide chains in the advanced enzyme structure are compact due to extensive folding, with most of the hydrophilic amino acid groups being located on the exterior surface and the hydrophobic groups inside.

There are various functional groups in enzyme molecules, such as $-NH_3$, $-COOH$, $-SH$, and $-OH$, with varying functional characteristics. Only some of the specific functional groups are directly associated with the catalytic activity of enzymes, such as imidazolyl groups on histidine, hydroxyl groups on serine, sulfhydryl groups on cysteine, and the carboxyl groups on glutamate and aspartic. The active sites of enzymes are composed of binding sites and catalytic sites. Binding sites form bonds with substrates and determine the specificity of enzymes, while the catalytic sites act as the catalyst. The functional groups and conformation of the active sites vary according to the different enzymes. For a single enzyme (the enzyme with the only component of amino acid), the active sites are the few relatively close amino acid residue groups within the spatial structure of the enzyme molecule or the side chain groups on these residues. These residues may be relatively far apart in the primary structure or even within the different peptide chains; however, they can be in close proximity in the spatial conformation. For conjugated enzymes (where the enzyme needs a coenzyme or prosthetic group to participate in catalysis), the active sites also contain certain chemical structures of coenzymes or prosthetic groups besides the amino acid residue groups composing the active site. Prosthetic groups are mostly inorganic ions or small molecular organic compounds, which may be the component of the active sites themselves. Prosthetic groups act as the bridge between substrates and enzymes, stabilizing the catalytic activity of apoenzymes and the essential molecular conformation, and acting as a carrier for the transfer of hydrogen atoms, electrons, some special atoms, and groups for the catalytic reactions.

There is generally one active site in an enzyme molecule. However, some enzyme molecules are composed of multiple subunits and may contain several active sites. There is often only one catalytic center, which generally contains two or three amino acid residual group side chains. The number of binding centers varies from one to several with different substrates and reaction properties. Further, due to the different substrate properties, the number of amino acids contained in the various binding centers varies. Among the enzymes of a known structure, the active sites usually exist within the fissure structure of the enzyme molecules. The hydrophobic structure of the fissure is beneficial to the binding of enzymes and substrates and avoids the entry of water molecules. However, if water molecules also take part in the reaction, polar residue groups should be present in the fissure to construct the required special microenvironment for catalysis.

The specificity of the binding of enzymes and substrates depends on the special spatial arrangement of the related atoms in the active centers, and substrates must be structurally matched with the active center. Additionally, the active center structure will change when binding with substrates, forming a complementary structure with the substrate via the inducing effect.

It should be noted that the formation of active sites requires the apoenzyme molecules with a certain spatial conformation. The amino acid residual groups in enzymatic molecules, except for those at the active sites, are necessary to stabilize the spatial conformation of enzyme molecules. Thus, they play an important role in the catalytic activity of enzymes. For example, amino acid residual groups are related to the regulation of enzymatic activity, the formation of the correct spatial structure of enzyme molecules, and other properties such as immunity of enzyme molecules.

9.2.2 *Catalytic mechanisms of biological enzymes*

The catalytic mechanisms of enzymes mainly include the approaching and orientation effect, conformation variation effect, acid-alkaline catalysis, covalent catalysis, microenvironment effect, self-shearing, and self-splicing.

9.2.2.1 *Approaching and orientation effect*

All enzyme molecules have an active center. The groups at the active center approach the substrates making the substrate and catalytic groups on the enzyme active centers follow the geometric orientation with the correct direction, favoring the formation of intermediate products and the progress of catalytic reactions.

The approaching and orientation effect between enzymes and substrates has the following impact on catalytic reactions:

(1) The concentration of the substrate molecules near the enzyme active center increases, thus increasing the reaction rate.
(2) The groups at the enzyme active center show an orbital guidance effect on the substrate molecules, thus decreasing the required reaction activation energy.
(3) Intermediate product ES is formed in the reaction of enzyme and substrates, making the intermolecular reaction become an intramolecular reaction, thus increasing the reaction rate. Additionally, the ES life span is 10^{-7} to 10^{-4} s, while the mean life span of the combination of the two molecules via random collision is 10^{-13} s. Thus, the former is 10^6–10^9 times longer than the latter, making the probability of the enzymatic catalytic reaction increase with higher reaction efficiency.

9.2.2.2 *Conformation variation effect* [11, 12]

As the enzyme molecules approach the substrate molecules, conformation variations occur in both the enzyme and substrate molecules due to their interaction. This is beneficial to the binding and reaction between the enzymes and substrates, greatly increasing the reaction rate.

(1) The substrate induces the conformation variation of the enzyme molecules. In 1958, Koshland proposed the Induced Fit Theory. He considered that the conformation of the enzyme molecules was not invariable. The enzyme molecules are induced by the substrates when they are close, leading to conformation variation to facilitate the binding of enzyme molecules with substrates.

(2) The variation in substrate conformation is induced by enzyme molecules. When enzyme molecules are in near proximity to substrates, substrates show conformation variations with various distorted deformation under the induction of enzyme molecules in order to combine with the enzyme active center. These are termed the transition state structures, greatly decreasing the required activation energy in the reaction and accelerating the reaction rate.

9.2.2.3 *Acid-Alkaline catalysis mechanism*

The enzymatic catalysis mechanism is the proton transferring action between the enzyme and substrate molecules in acid-alkaline catalysis, decreasing the required activation energy in the reaction and accelerating the reaction. The acid and alkaline catalysis groups of apoenzymes are provided by the amino acid residual group side chains, with imidazolyl being the most common group. Under physiological conditions, imidazolyl can be used as both the proton donor and proton acceptor, and the rates of proton accepting and releasing are both fast and equal to each other.

As the proton donor in apoenzymes, the acid catalysis group is a conjugate acid, while the alkaline catalysis group is a conjugate alkaline. The catalysis of conjugate acid can be described as forming a hydrogen bond with an oxygen atom on the carbonyl group of the substrate and making the carbon atom on the carbonyl group carry a greater positive charge, which easily attracts the lone pairs on water molecules, decreasing the activation energy of the covalent bond between the carbonyl group and water molecule. Subsequently, the conjugate acid transfers H^+ to the carbonyl oxygen on the substrate and becomes a conjugate alkaline itself. Then, the conjugate alkaline immediately absorbs H^+ from the water molecules and returns to being a conjugate acid, finishing the catalysis process. The catalysis of conjugate alkaline can be described as the formation of hydrogen bonds with the hydrogen atom on the water molecule, increasing the electronegativity of the oxygen on the water molecule. This facilitates the nucleophilic attack on the carbon atom in the carbonyl group and thus decreases the activation energy of the carbon–oxygen bond, accelerating the reaction rate. The conjugate alkaline attracts the H^+ from the water molecule and becomes a conjugate acid. Subsequently, the conjugate acid transfers the attracted H^+ to the substrate groups and returns to being a conjugate alkaline, finishing the catalysis process.

9.2.2.4 *Covalent catalysis mechanism* [13]

In the catalytic process, an intermediate compound is formed from the reaction of enzymes and substrates. This intermediate compound is the covalent intermediate product formed from the attack of certain groups of substrates by certain enzymes. According to the substrate group being attacked by the enzymes, covalent catalysis can be either nucleophilic or electrophilic catalysis.

Nucleophilic catalysis involves the attack by electron-rich groups (nucleophilic groups) within the enzyme molecules on electron-deficient groups (electrophilic groups) in substrates to form a covalent intermediate product. Common nucleophilic catalytic reactions include nucleophilic substitution and nucleophilic addition reactions. Electrophilic catalysis is the process through which electron-deficient groups (electrophilic groups) within the enzyme molecules attack the electron-rich groups (nucleophilic groups) in substrates to form a covalent electrophilic intermediate product.

9.2.2.5 *Microenvironment effect*

The microenvironment refers to a special hydrophobic reaction environment where the catalytic groups of the enzyme active center are located. This special hydrophobic reaction environment affects the binding between enzymes and substrates and influences the dissociation of catalytic groups, accelerating the reaction process, thus being called the microenvironment effect.

In addition, there are self-shearing mechanism and self-splicing mechanism to explain the effects of catalytic activity of enzyme.

9.3 Preparation principle of CD-based enzyme mimics

Enzyme mimics are nonprotein molecules that possess similar functions as natural enzymes but with simpler structures and which can be used as models to simulate the binding of enzymes and substrates to enhance the activity and stereoselectivity of reactions. The study of enzyme mimics is an important component of the supramolecular chemistry [14].

CD has a unique conical molecular structure, and its primary and secondary hydroxyl groups are respectively located at the small and large open ends of the cylinder, leading to the hydrophilic nature of the external surface [15]. The inner cavity is composed of hydrogen atoms on C-3 and C-5 and glycoside oxygen atoms and is therefore hydrophobic. Thus, CD can combine with various molecules to form a supramolecular inclusion compound. CD is an excellent subject in the field of molecule recognition, but also a good seminatural receptor in the research of enzyme mimics. With selective modification, the structure (extension in the hydrophobic region) and properties (introduction of functional groups) of CD derivatives can be further improved. Thus, CD derivatives are suitable for use as the main materials in the construction of enzyme models [1, 16].

In summary, CD is an ideal enzyme model, as described below:

(1) It has good water solubility and can selectively bond with the substrate in an aqueous solution. The primary and secondary hydroxyl groups dimensionally located at both sides of the cavity can participate in the reaction with the substrate.
(2) The primary and secondary hydroxyl groups have different proton donor activities and handle the transfer system between catalysts and protons through a chemical method.

(3) The internal cavity of the molecule is composed of C_3–H, C_5–H, and 1,4-glucoside bond oxygen of glucose group, forming a relatively rigid hydrophobic cavity that favors the selective binding and positioning of organic molecules.

CD can catalyze chemical reactions, such as hydrolysis reaction of ester, alkylation of halogen, and Diels–Alder reactions, due to its hydrophobic cavity, which is similar to that of natural enzymes [17–20]. However, the binding constant of CD and substrate is usually 10^4 mol/L, which is smaller than the binding constant of the enzyme and substrate. Additionally, the catalytic ability of CD hydroxyls is limited, and thus, the construction of CD-based enzyme mimics mainly focuses on the modification of CD, mainly through the introduction of molecular recognition or catalytic functional groups that can improve its hydrophobic binding and catalytic function [21].

Generally, the construction of CD-based enzyme mimics needs to consider the characteristics of CD structures, including the spatial allocation of substrate and CD, the synergic effect of the catalysts, the construction of the microenvironment, the selection of the binding sites and binding modes, the assistance of the metal ions, and the construction of dual binding sites [22, 23]. The cavity binding ability of CD and the catalytic groups must be effectively combined. Further, these groups must be ingeniously configured to maximize their correlation. Generally, substrate molecules with larger and more complex structures will be more favorable to the generation of binding sites with appropriate energy. As a result, multipoint binding is an effective design plan.

Modification of the main and secondary sides of CD is the most important factor to consider when constructing an enzyme model. Studies found that the introduction of catalytic groups to either side was effective [24–28]. However, certain substrates preferred the secondary side (with the larger opening), and the introduction of the catalytic groups on the secondary side under this condition showed better activity [26, 29–31]. If the two functional groups are introduced on the same side of CD, their synergistic effect could form a new system to improve catalytic activity. Additionally, since the two groups participate in the catalytic reactions synergistically, the electrostatic environment of the CD-binding interface can be kept within the required range. For example, introducing 1,4-dihydro nicotinamide (NAH) on the main side of CD led to a fiftyfold higher rate of reduction of ninhydrin compared to that of NADH, while introduction of NAH on the secondary side led to a sixtyfold increase in the reduction rate compared to that of NADH. When two NAH groups were introduced on the main side of CD, the reduction rate of ninhydrin was improved hundredfold compared to that of NADH [32–34].

The introduction of two imidazole groups on the main side of CD was able to stimulate ribonucleic acid (RNA) [35]. The K_{cat} of the catalytic hydrolysis of cyclic phosphate by the enzyme mimic was 120×10^{-5} s^{-1}, which was much higher than that of the nonenzymatic reaction ($K_{uncat} 1 \times 10^{-5}$ s^{-1}). Moreover, the enzyme mimic was able to manifest good stereoselectivity, and the ratio of the two phosphates generated was 99:1, while the ratio of products from the nonenzymatic reaction of cyclic phosphates in aqueous NaOH solution was 1:1 (Fig. 9.1). Isotopic and kinetic studies showed that two imidazolyl groups showed a synergistic effect in the catalysis and acted as the acid and alkaline catalysts, respectively. In addition, the relative location of the two imidazolyls on β-CD was also important for catalysis, and the enzyme mimic showed a high catalytic efficiency and stereoselectivity only when the two imidazolyls were adjacent to each other.

Introducing groups with a special conformation into CD could lead to enzyme mimic models with stereoselectivity. Pyridoxamine

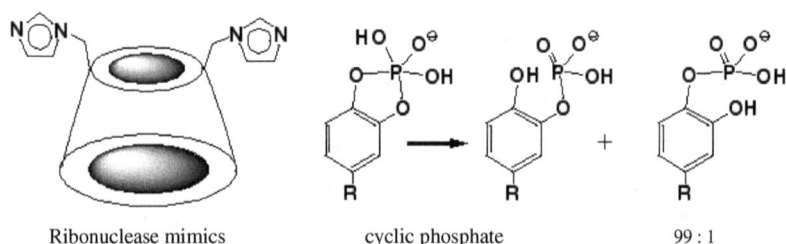

| Ribonuclease mimics | cyclic phosphate | 99 : 1 |

Fig. 9.1.   Catalytic hydrolysis of cyclic phosphate by CD-based enzyme mimics.

bonded to the primary hydroxyl group of β-CD led to the catalysis of the amino conversion reaction of α-ketonic acid [8, 36, 37]. The reaction rate was hundredfold faster than that using pyridoxamine alone as a catalyst, and the catalysis showed stereoselectivity during the process. Pyridoxamine was introduced to the main side of β-CD via a double chain to obtain a transaminase model. The results showed that, when the pyridoxamine group was located at one side of the CD, it selected preferential bonding with the substrate of metasubstituted phenyl acetylformic acid. Conversely, when it was located above the CD cavity, it selectively bonded with the substrate of parasubstituted phenyl acetylformic acid.

In addition, when the CD units are bridged with each other via some functional groups and the two adjacent CD cavities can synergistically participate in the interaction of inclusion complexation with the target molecules of appropriate shape and size, a supramolecular complex with increased stability and better ability to simulate bioenzymes is formed. Based on this, a series of CD polymers were synthesized [38–43].

Metal ions can play an important role in many electron-transfer systems, being often used as "superacid" catalysts with a directional function. Therefore, the introduction of metal ions into enzyme mimic models can improve their catalytic efficiency. The earliest CD metal complex formed was the peptidase model, formed by the complexation of one or two oxime-modified CDs with Cu(II) and Ni(II) ions [44, 45]. This complex utilizes metal ions to catalyze the hydrolysis reaction of the substrate included in the CD cavity.

CD-based enzyme mimics have potential applications in the field of bionic sensors. For example, nicotine groups were introduced on the secondary side of β-CD to simulate dehydrogenase, and this was applied in a bionic sensor. Using electrochemistry and fluorescence techniques, the ethanol, acetone, dopamine, and propanol reaction conditions of the CD inclusion ability and the nicotine group electron-transfer ability in the conditions of anion mixed with hydrophilic and hydrophobic were observed. It was found that the size and hydrophobicity of the substrates directly influenced the catalytic activity of the enzyme models. The prepared sensors were also recyclable.

9.4 Typical models of CD-based enzyme mimics

9.4.1 *Hydrolase mimic based on CD*

Currently, the study of enzyme mimics based on a CD backbone has made great progress, with several biochemical reactions being artificially simulated. Among these, metal hydrolase is an enzyme mimic with extensive recent applications. An early enzyme model was the hydrolase mimic of *p*-nitrophenyl acetate reported in 1970 (Fig. 9.2(a)) [44]. This model was constructed by the introduction of imidazolyl to the primary or secondary hydroxyl groups on CD. When one imidazolyl group was used to directly modify the primary hydroxyl of β-CD, the catalytic activity of the model was relatively weak, whereas the catalytic activity of CD modified on the secondary hydroxyl group was thousandfold higher at pH of 7.5. The catalytic activity ratio of the two modified CDs was larger than 70, showing an obvious location effect. The great difference in catalytic activity was due to the different orientations between the active groups and reaction center (carbonyl carbon in the substrate) of the two CDs. The substrate entered the cavity with the secondary hydroxyl group as head, which was more appropriate to the modification of CD by the secondary hydroxyl group.

When two imidazolyl groups were introduced on one side (first side) of the β-CD primary hydroxyl group, a charge-substitution system was formed due to the synergistic effect of the two groups, leading to a selective catalytic hydrolysis effect. There were three

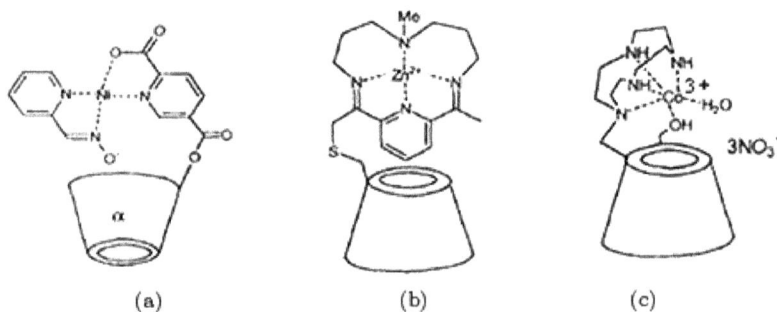

(a)   (b)   (c)

Fig. 9.2.   Metal hydrolase model based on CD: (a) Model 1, (b) model 2, (c) model 3.

bonding modes of the imidazolyl groups in the model: AB, AC, and AD (according to the glucose order). At pH 7, only the AB isomer led to better catalysis on the hydrolysis of phosphoric acid *tert*-butyl ester, with a hydrolysis effect of high stereoselectivity. Analyzing the geometrical construction of the AB isomer, it can be considered that the imidazolyl group was used as the acid catalyst. The protonation of phosphate anions made the P–O on the substrate break preferentially, and the water molecules were transferred by the neutral imidazolyl group. As a consequence, the intermediate was transformed into the product without quasirotation. Additionally, the catalyst underwent no change and continued to react, increasing the turnover of the reaction. The rate constant of catalytic hydrolysis of the cyclic phosphate was not high using this enzyme mimic model, but it possessed typical enzyme-like activity.

The zinc enzyme model subsequently established (Fig. 9.2(b)) was able to effectively promote the hydrolysis of aryl phosphate [46]. Another good hydrolase model of tetraaza macrocycle cobalt complex (Fig. 9.2(c)) is one of the most effective acyl transfer catalysts [47]. The enzyme model was obtained by the covalent bonding of the complex with β-CD, increasing the hydrolysis rate of nitrophenyl acetate by 900-fold. The enzyme model obtained by bonding the same substituent group with the secondary hydroxyl group of CD led to improvement of the hydrolysis rate of nitrophenyl acetate by 2900- to 3700-fold.

The catalytic activity of the artificial enzyme, β-benzyme, synthesized in 1985, is in the same order of magnitude as that of natural chymotrypsin and with a high stability. This model utilized CD as the substrate-binding site. The carboxyl group, imidazolyl group, and a secondary hydroxyl group of CD composed the catalytic active center, which was similar to that of natural enzymes, realizing the full simulation of chymotrypsin (Fig. 9.3) [9]. The hydrolysis rate of *tert*-butyl phenyl acetate catalyzed by β-benzyme was above onefold faster than that of *p*-NPAC catalyzed by natural enzyme, with an equivalent K_{cat}/K_m as that of natural enzymes.

An artificial supramolecular nanozyme based on β-CD-modified gold nanoparticles shown strong hydrolase activities for catalyzing the

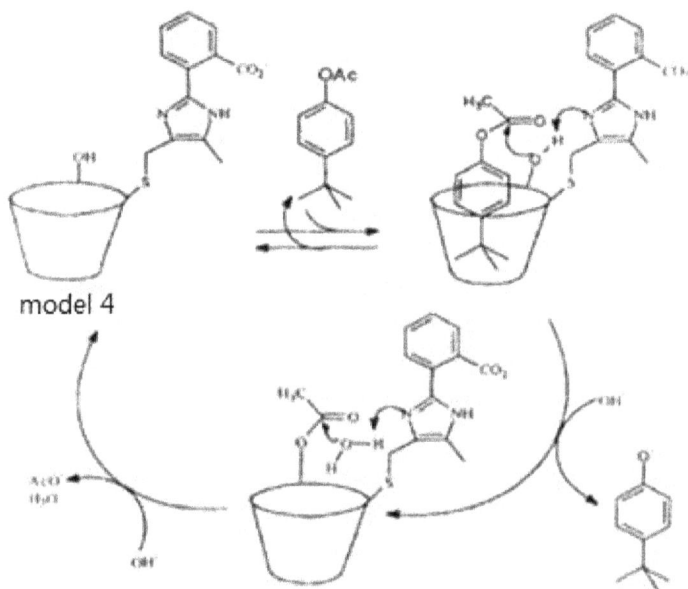

Fig. 9.3.   Model of an artificial enzyme mimic of chymotrypsin (β-benzyme) and the catalysis mechanism.

cleavage of an active ester 4,4'-dinitrodiphenyl carbonate (DNDPC) [48]. For mimicking catalytic action of natural hydrolytic metalloenzymes containing two or more metal center in their active sites, the synergistic effect of two metal ions and the microenvironment of enzyme catalysis, an artificial nanozyme model was developed by the supramolecular complexation of a β-CD-modified gold nanoparticle and metal catalytic centers (Fig. 9.4). The CD-based monolayer was constructed on the surface of gold nanoparticle by using thiol-modified CD. The CD-modified gold nanoparticle was utilized as a backbone to construct a supramolecular artificial enzyme. The catalytic behaviors of the β-CD-modified gold nanoparticles with adjacent multimetal catalytic centers were investigated as an esterase mimic. The study shows that CD-modified gold nanoparticles are suitable building blocks for the design of effective enzyme models. The kinetic analyses indicated that the synergic action of multimetal catalytic centers and three-dimensional structure of Au nanoparticle may be the possibility for rate enhancement of carbonate hydrolysis.

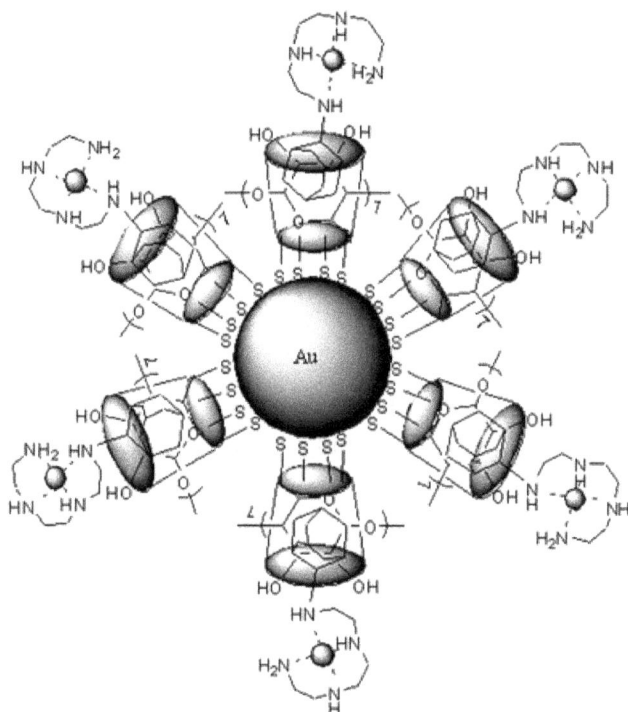

Fig. 9.4. Structure of the CD-based nanozyme model on a gold nanoparticle.

Furthermore, the exploration of the biochemical ability of functional gold nanoparticles will expand the application of nanoparticles in biological approach.

9.4.2 *Artificial transaminase*

Another example of the synergic effect of introducing two active groups is the aminopherase model. Natural aminopherase usually requires the activation of zymogen, firstly by the coenzyme pyridoxic acid (vitamin B_6), for the catalysis of various amino acid reactions such as decarboxylation, elimination, and transaminase reactions. The enzyme mimic model (Fig. 9.5) formed by bonding pyridoxamine with the primary hydroxyl group of CD was used to simulate the transaminase reactions [40, 49].

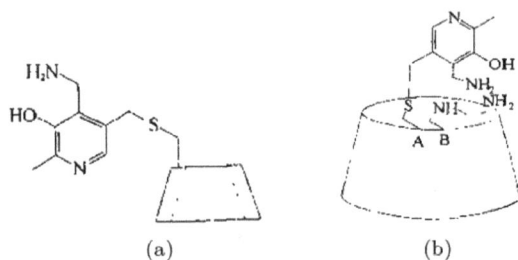

Fig. 9.5.   Artificial transaminase model: (a) Transminase model I, (b) transaminase model II.

Enzyme mimic I was able to increase the transformation rate of indolepyruvic acid to indole alanine by 200-fold compared to that of pyridoxamine itself. This made the excess percentage of L-tryptophan formed from indolepyruvic acid and L-alanine formed from phenyl-pyruvic acid antimers reach 12% and 67%, respectively, indicating that the reaction showed an obvious optical induction effect.

Studies showed that Schiff alkaline was formed by the reaction between aldehyde group and lysine residual group of the ε-azyl group at the natural enzyme active centers relying on vitamin B_6. Under various noncovalent-bonding reactions, the coenzymes were fixed in the zymo proteins and three bonding around the α-carbon in amino acid of the transition state intermediate were effectively decomposed with the aid of the electron-sag effect of the conjugated pyridyl group. Based on this, new aminopherase model structures were developed. The AB mode of sulfonic acid ester formed by the reaction between 1,3-benzenedisulfonyl chloride and β-CD was used as the interme-diate. After the transformation to diiodosalicylic CD, the enzyme mimic model II was obtained by the reaction with pyridoxamine and ethane diamine in two steps. This model had a high chiral selectivity to the transaminase reaction of ketonic acid, with the formed L-tryptophan of 95%. The *L/D* ratio of aromatic L-amino acid produced by catalysis is listed in Table 9.1. The planar intermediate C...N...C was formed in the reaction, and diaminoethane, acting as the proton transfer catalyst, was fixed at one side of the plane, which made the reactants highly enantioselective. Thus, if the microenvironment provided by the protein enzyme is not considered, it is impossible

Table 9.1.   Optical activity of the transformation of
α-ketonic acid to α-amino acid.

α-ketonic acid	α-amino acid	L/D
R = Ph-CH$_2$–	Ph-CH$_2$-CH-CO$_2$H $\quad\quad$ NH$_2$	98/2
CH$_3$–	CH$_3$-CH-CO$_2$H $\quad\quad$ NH$_2$	98/2
		95/5

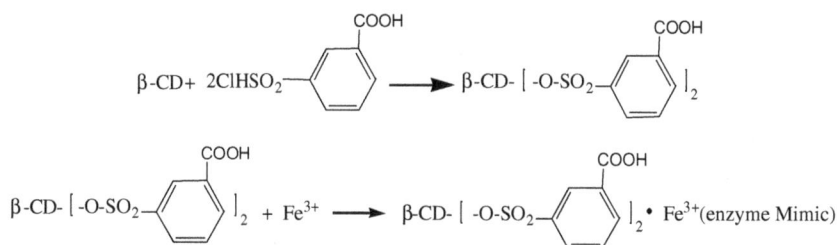

Fig. 9.6.   Preparation reaction of amino-acid oxidase enzyme mimics.

to achieve a high reaction rate and high substrate specificity and
selectivity.

9.4.3 *Constrsuction of oxidoreductase mimic based on CD*

In recent years, a series of CD-based oxidoreductase models were
synthesized [32–34, 50]. For example, an amino acid oxidase model
was constructed by a complex prepared from CD carbonyl benzenesulfonyl chloride and iron trichloride (Fig. 9.6). This enzyme was
used to produce phenyl-pyruvic acid from the oxidative deamination
of phenylalanine, and phenylacetic acid was further prepared by the
oxidative decarboxylation of phenyl-pyruvic acid in the presence of
excess hydrogen peroxide. If the reaction was carried out under the

$$\beta\text{-CD} + 2\text{ClHSO}_2\text{—}\underset{\substack{\text{COOH}}}{\bigcirc} \longrightarrow \beta\text{-CD}\text{—}[\text{O-SO}_2\text{—}\underset{\substack{\text{COOH}}}{\bigcirc}]_2$$

[Abbreviation: β-CD-(M)$_2$]

$$\beta\text{-CD-(M)}_2 + \text{Fe}^{3+} \longrightarrow \beta\text{-CD-(M)}_2 \bullet \text{Fe}^{3+}$$

[Abbreviation: enzyme Mimic]

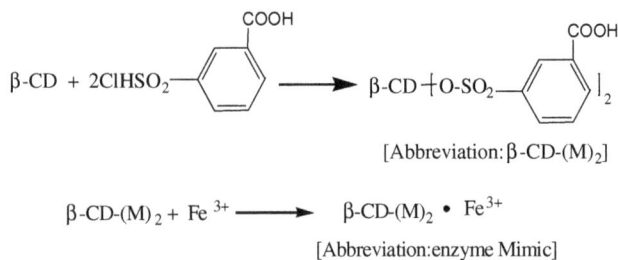

Fig. 9.7.   Preparation reaction of glucose oxidase enzyme mimic.

weak basic condition of dimethylformamide, tyrosine was formed by the oxidation of phenylalanine.

Glucose oxidase mimics can be constructed by the complex prepared from β-CD carbonyl benzenesulfonyl chloride and iron trichloride (Fig. 9.7). The enzyme–substrate complex was composed of this enzyme mimic and hydrogen peroxide, which could be used to catalyze the oxidation of glucose and produce gluconic acid. The catalytic kinetics of enzyme mimics was characterized by the characteristic absorption (at 365 nm) of the product obtained from gluconic acid and o-toluidine, showing that the catalytic efficiency was much higher than that of the inorganic catalyst ($\text{FeCl}_3 + \text{H}_2\text{O}_2$).

Agricultural by-products, such as peanut shells, corncob, wheat-straw, cotton hull, and bagasse, can be used as raw materials and be treated by dilute sulfuric acid (3–5%), leading to the production of furfural by the dehydration of pentose, followed by the formation of furancarboxylic acid by further oxidation of furfural. Furancar-boxylic acid is an important industrial raw material for plasthetics, medicines, and pesticides, with an important economic value. Several approaches are available for the preparation of furancarboxylic acid by furfural oxidation, including the Cannizzaro method, potassium permanganate, potassium dichromate, sodium hypochlorite, hydro-gen peroxide, and pure oxygen methods. However, these chemical methods produce a large amount of by-products and the catalysts adopted in some methods are toxic and expensive. Using β-CD for the construction of a new catalyst, furfural can be transformed into furancarboxylic acid in 1 h, in acidic conditions at 50°C. The catalytic

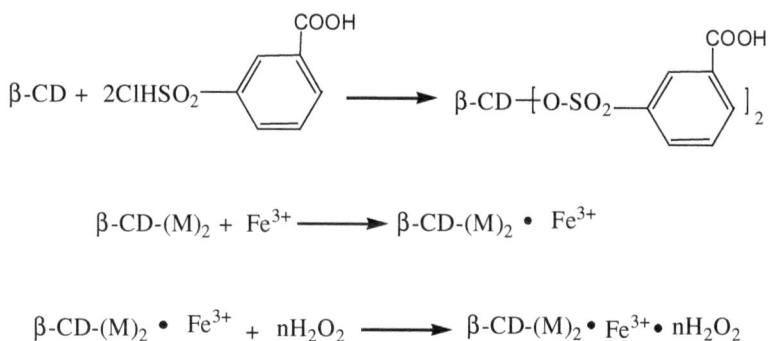

$$\beta\text{-CD} + 2\text{ClHSO}_2\text{—}\underset{}{\bigcirc}^{\text{COOH}} \longrightarrow \beta\text{-CD}\text{–}[\text{O-SO}_2\text{—}\underset{}{\bigcirc}^{\text{COOH}}]_2$$

$$\beta\text{-CD-(M)}_2 + \text{Fe}^{3+} \longrightarrow \beta\text{-CD-(M)}_2 \bullet \text{Fe}^{3+}$$

$$\beta\text{-CD-(M)}_2 \bullet \text{Fe}^{3+} + n\text{H}_2\text{O}_2 \longrightarrow \beta\text{-CD-(M)}_2 \bullet \text{Fe}^{3+} \bullet n\text{H}_2\text{O}_2$$

Fig. 9.8.   Preparation reaction of CD-based enzyme mimic complex.

rate of this method is 2.7×10^4-fold faster than that of H_2O_2 and 64-fold that of $FeCl_3 + H_2O_2$, with a reactant conversion rate of 97%. The catalyst is precipitated in organic solvents such as acetone and is separated from the product of furancarboxylic acid, making it reusable. Therefore, this method has the advantages of easy cycle regeneration, low cost, no toxicity, convenient usage, high catalytic efficiency, and short reaction time.

The preparation of the CD-based enzyme mimic complex is shown in Fig. 9.8.

In addition, a CD-based enzyme mimic that can selectively oxidize substrates with a specific spatial conformation has been prepared. At present, the prepared CD-based enzyme mimics can not only be used for the catalysis of chemical reactions but can also provide the possible spatial conformations for a transition complex.

In recent years, studies on enzyme mimics using bridged CDs have achieved great progress. Since the functional groups of the two CDs and bridge groups constitute the catalytic activity center with cooperative inclusion and multiple recognition functions, bridged CDs can better simulate the recognition and catalysis functions of enzymes to substrates. For example, a CD dimer bridged by tri-amino ethylamine can be used for the hydrolysis of *P*-nitrophenyl carbonate, with a catalytic efficiency 150-fold higher than that of the blank reaction.

Breslow performed extensive research on the bridged CD in recent years, not only reporting the synthesis method and binding ability of a

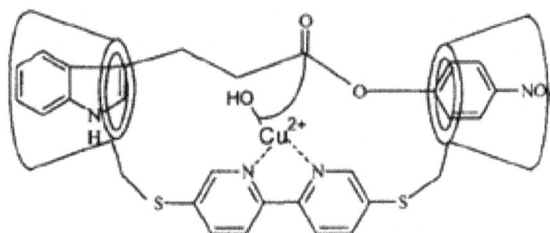

Fig. 9.9.   Hydrolysis mechanism of esters by bridged CD in the presence of copper ions.

series of bridged CDs but also successfully utilizing enzyme mimics in the catalytic hydrolysis reaction of dual hydrophobic organic esters. When the organic ester substrates with dual hydrophobic sites are enclosed in the two CD cavities, the divalent copper-ion complex at the bridge group was located near the ester group in the substrate, making it beneficial to attack the –OH group in esters and obviously accelerate the hydrolysis reaction. The catalytic rate was improved by 1.8×10^4-fold compared with that of the nonenzymatic reaction (Fig. 9.9).

Similarly, glutathione peroxidase (GPx) can be simulated by using the artificially synthesized 6-selenium-bridged β-CD. GPx can use reduced glutathione as the substrate to eliminate free radicals *in vivo* and avoid lipid peroxidation through the catalytic reduction of hydrogen peroxide, which had an obvious effect on treating and preventing cardiovascular diseases and cancers. The study utilized the special structure of β-CD, introducing the active group –Se of GPx into the bridge chain and synthesizing a bridged CD. By comparing the activity of the Se-bridged β-CD mimic and other small molecular enzyme mimics (e.g., ebselen(2-phenyl-1,2-benzisoselenazol-3(2H)-ketone)), it was found that Se-bridged β-CD showed a higher catalytic activity for GPx due to the increase in binding sites using the CD cavity (Fig. 9.10).

For example, the GPx mimic synthesized by 2- or 6-Se β-CD showed a good catalytic activity of GPx, with a biological activity over 4.3-fold that of the famous GPx mimic Ebselen. Studies also found that substitution of the 2- or 6-site had an obvious influence on the activity and that an increase in the catalytic groups could enhance the

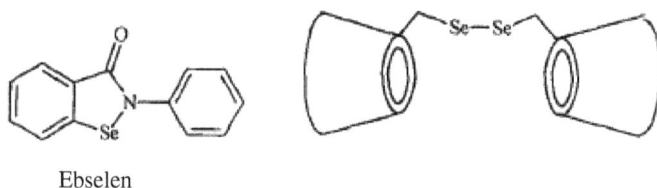

Ebselen

Fig. 9.10.　Glutathione peroxidase mimic.

reaction activity. If Se atoms were introduced into the 2- or 6-site of β-CD, the new CD mimics manifested an even higher catalytic activity of GPx, and the activity of 2-Se-based β-CD was 47-fold higher than that of Ebselen.

Besides hydrolase, oxidoreductase, ribonuclease, and transaminase, CD can also be used to construct various enzyme mimics, such as carbonic anhydrase, thiaminase, aldol condensation enzyme, and biotin, which have obtained good results.

References

1. Breslow R, Dong SD. (1998). Biomimetic reactions catalyzed by cyclodextrins and their derivatives. *Chemical Reviews*, 98, pp. 1997–2011.
2. D'Souza VT. (2003). Modification of cyclodextrins for use as artificial enzymes. *Supramolecular Chemistry*, 15(3), pp. 221–229.
3. Motherwell WB, Bingham MJ, Six Y. (2001). Recent progress in the design and synthesis of artificial enzymes. *Tetrahedron*, 57, pp. 4663–4686.
4. Stevenson JD, Thomas NR. (2000). Catalytic antibodies and other biomimetic catalysts. *Natural Product Reports*, 17, pp. 535–577.
5. Wulff G, Gross T, Schonfeld R. (1997). Enzyme models based on molecularly imprinting polymers with strong esterase activity. *Angewandte Chemie International Edition*, 36, pp. 1962–1964.
6. Dugas H, Penney C. (1981). *Bioorganic Chemistry*. New York: Springer.
7. Hapiot F, Ponchel A, Tilloy S, Monflier E. (2011). Cyclodextrins and their applications in aqueous-phase metal-catalyzed reactions. *Comptes Rendus Chimie*, 14(2–3), pp. 149–166.
8. Breslow R, Canary JW, Varney M, Waddell ST, Yang D. (1990). Artificial transaminases linking pyridoxamine to binding cavities — controlling the geometry. *Journal of the American Chemical Society*, 112(13), pp. 5212–5219.
9. D'Souza VT, Bender ML. (1987). Miniature organic models of enzymes. *Accounts of Chemical Research*, 20, pp. 146–152.
10. Reslow R. (1995). Biomimetic chemistry and artificial enzymes: catalysis by design. *Accounts of Chemical Research*, 28, pp. 146–153.

11. Koshland DEJ. (1963). Correlation of structure and function in enzyme action. *Science*, 142, pp. 1533–1541.
12. Wolfenden R. (1969). Transition state analogues for enzyme catalysis. *Nature* 223, pp. 704–705.
13. Freeman AFWH. (1985). *Enzyme Structure and Mechanism*, second edition. New York: FEBS (Gutfreund H (Ed.)).
14. Lehn J-M. (1995). *Supramolecular Chemistry: Concepts and Perspectives.* Weinheim: VCH.
15. Villiers A. (1891). Discovery of cyclodextrins. *Comptes Rendus de l'Académie des Sciences*, 112, pp. 536–538.
16. Rekharsky MV, Inoue Y. (1998). Complexation thermodynamics of cyclodextrins. *Chemical Reviews*, 98(5), pp. 1875–1917.
17. Kumar VP, Reddy MS, Narender M, Surendra K, Nageswar YVD, Rao KR. (2006). Aqueous phase mono-protection of amines and amino acids as N-benzyloxycarbonyl derivatives in the presence of beta-cyclodextrin. *Tetrahedron Letters*, 47(36), pp. 6393–6396.
18. Park JB, Kwon CHD, Lee KW, Choi GS, Kim DJ, Seo JM, Kim SJ, Joh JW, Lee SK. (2008). Hepatitis B virus vaccine switch program for prevention of *per se* hepatitis B virus infection in pediatric patients. *Transplant International*, 21(5), p. 510.
19. Reddy MA, Bhanumathi N, Rao KR. (2002a). A mild and efficient biomimetic synthesis of alpha-hydroxymethylarylketones from oxiranes in the presence of beta-cyclodextrin and NBS in water. *Tetrahedron Letters*, 43(17), pp. 3237–3238.
20. Reddy MA, Reddy LR, Bhanumathi N, Rao KR. (2002b). A novel and efficient biomimetic hydrolysis of oxiranes to 1,2-diols catalysed by beta-cyclodextrin in water under neutral conditions. *Organic Preparations and Procedures International*, 34(5), pp. 537–540.
21. Khan AR, Forgo P, Stine KJ, D'Souza VT. (1998). Methods for selective modifications of cyclodextrins. *Chemical Reviews*, 98(1998), pp. 1977–1996.
22. Marinescu L, Bols M. (2010). Cyclodextrins as supramolecular organo-catalysts. *Current Organic Chemistry*, 14(13), pp. 1380–1398.
23. Zhao W, Zhong Q. (2012). Recent advance of cyclodextrins as nanoreactors in various organic reactions: a brief overview. *Journal of Inclusion Phenomena and Macrocyclic Chemistry*, 72(1–2), pp. 1–14.
24. Coleman AW, Zhang P, Parrot-Lopez H, Ling C-C, Miocque M, Mascrier L. (1991). The first selective per-tosylation of the secondary OH-2 of β-cyclodextrin. *Tetrahedron Letters*, 32, pp. 3997–3998.
25. Hanessian S, Benalil A, Laferriere C. (1995). The synthesis of functionalized cyclodextrins as scaffolds and templates for molecular diversity, catalysis, and inclusion phenomena. *Journal of Organic Chemistry*, 60, pp. 4786–4797.
26. Maletic M, Wennemers H, McDonald DQ, Breslow R, Still WC. (1996). Selective binding of the dipeptides L-Phe-D-Pro and D-Phe-L-Pro to β-cyclodextrin. *Angewandte Chemie, International Edition*, 35, pp. 1490–1492.

27. Siegel B, Pinter A, Breslow R. (1977). Synthesis of cycloheptaamylose 2-, 3-, and 6-phosphoric acids, and a comparative study of their effectiveness as general acid or general base catalysts with bound substrates. *Journal of the American Chemical Society*, 99, pp. 2309–2312.

28. Ueno A, Breslow R. (1982). Selective sulfonation of a secondary hydroxyl group of β-cyclodextrin. *Tetrahedron Letters*, 23, pp. 3451–3454.

29. Hubbard BK, Beilstein LA, Heath CE, Abelt CJ. (1996). Synthesis and characterization of dicyanoanthracene-tethered β–cyclodextrins. *Journal of the Chemical Society, Perkin Transactions* 2, pp. 1005–1009.

30. Mortellaro MA, Hartmann WK, Nocera DG. (1996). Regioisomeric effects on the excited state processes of a cyclodextrin modified with a lumophore. *Angewandte Chemie, International Edition*, 35, pp. 1945–1946.

31. Yang Z, Breslow R. (1997). Very strong binding of lithocholic acid to β-cyclodextrin. *Tetrahedron Letters*, 38, pp. 6171–6172.

32. Kojima M, Toda F, Hattori K. (1980). The cyclodextrin-nicotinamide compound as a dehydrogenase model simulating apoenzyme-coenzyme-substrate ternary complex system. *Tetrahedron Letters*, 21(28), pp. 2721–2724.

33. Tabushi I, Kodera M. (1987). Flavocyclodextrin as a promising flavoprotein model. Efficient electron transfer catalysis by flavocyclodextrin. *Journal of the American Chemical Society*, 109(15), pp. 4734–4735.

34. Yoon C-J, Ikeda H, Kojin R, Ikeda T, Toda F. (1986). Reduction of ninhydrin with cyclodextrin-1,4-dihydronicotinamides as NADH models. *Journal of the Chemical Society, Chemical Communications* 14, pp. 1080–1081.

35. Breslow R, Doherty JB, Guillot G, Lipsey C. (1978). beta.-Cyclodextrinylbisimidazole, a model for ribonuclease. *Journal of the American Chemical Society*, 100(10), pp. 3227–3229.

36. Breslow R, Czarnik AW. (1983). Transaminations by pyridoxamine selectively attached at C-3 in .beta.-cyclodextrin. *Journal of the American Chemical Society*, 105(5), pp. 1390–1391.

37. Breslow R, Hammond M, Lauer M. (1980). Selective transamination and optical induction by a .beta.-cyclodextrin-pyridoxamine artificial enzyme. *Journal of the American Chemical Society*, 102(1), pp. 421–422.

38. Breslow R, Huang Y, Zhang XJ, Yang J. (1997a). An artificial cytochrome P450 that hydroxylates unactivated carbons with regio- and stereoselectivity and useful catalytic turnovers. *Proceedings of the National Academy of Sciences of the United States of America*, 94(21), pp. 11156–11158.

39. Breslow R, Zhang XJ, Huang Y. (1997b). Selective catalytic hydroxylation of a steroid by an artificial cytochrome P-450 enzyme. *Journal of the American Chemical Society*, 119(19), pp. 4535–4536.

40. Breslow R, Zhang XJ, Xu R, Maletic M, Merger R. (1996). Selective catalytic oxidation of substrates that bind to metalloporphyrin enzyme mimics carrying two or four cyclodextrin groups and related metallosalens. *Journal of the American Chemical Society*, 118(46), pp. 11678–11679.

41. Dong ZY, Liu JQ, Mao SZ, Huang X, Yang B, Ren XJ, Luo GM, Shen JC. (2004). Aryl thiol substrate 3-carboxy-4-nitrobenzenethiol strongly stimulating

thiol peroxidase activity of glutathione peroxidase mimic 2, 2'-ditellurobis(2-deoxy-beta-cyclodextrin). *Journal of the American Chemical Society*, 126(50), pp. 16395–16404.

42. Tastan P, Akkaya EU. (2000). A novel cyclodextrin homodimer with dual-mode substrate binding and esterase activity. *Journal of Molecular Catalysis A: Chemical*, 157(1–2), pp. 261–263.

43. Zhang BL, Breslow R. (1997). Ester hydrolysis by a catalytic cyclodextrin dimer enzyme mimic with a metallobipyridyl linking group. *Journal of the American Chemical Society*, 119(7), pp. 1676–1681.

44. Breslow R, Overman LE. (1970). "Artificial enzyme" combining a metal catalytic group and a hydrophobic binding cavity. *Journal of the American Chemical Society*, 92(4), pp. 1075–1077.

45. Sallas F, Marsura A, Petot V, Pinter I, Kovacs J, Jicsinszky L. (1998). Synthesis and study of new beta-cyclodextrin 'dimers' having a metal coordination center and carboxamide or urea linkers. *Helvetica Chimica Acta*, 81(4), pp. 632–645.

46. Breslow R, Singh S. (1988). Phosphate ester cleavage catalyzed by bifunctional zinc complexes: comments on the "p-nitrophenyl ester syndrome". *Bioorganic Chemistry*, 16(4), pp. 408–417.

47. Akkaya EU, Czarnik AW. (1988). Synthesis and reactivity of cobalt(III) complexes bearing primary- and secondary-side cyclodextrin binding sites. A tale of two CD's. *Journal of the American Chemical Society*, 110(25), pp. 8553–8554.

48. Li XQ, Qi ZH, Liang K, Bai XL, Xu JY, Liu JQ, Shen JC. (2008). An artificial supramolecular nanozyme based on beta-cyclodextrin-modified gold nanoparticles. *Catalysis Letters*, 124(3–4), pp. 413–417.

49. Tabushi I, Kuroda Y, Yamada M, Higashimura H, Breslow R. (1985). A-(modified B6)-B-[.omega.-amino(ethylamino)]-.beta.-cyclodextrin as an artificial B6 enzyme for chiral aminotransfer reaction. *Journal of the American Chemical Society*, 107(19), pp. 5545–5546.

50. Lawrence RM, Biller SA, Dickson JK, Logan JVH, Magnin DR, Sulsky RB, DiMarco JD, Gougoutas JZ, Beyer BD, Taylor SC, Lan S-J, Ciosek CP, Harrity TW, Jolibois KG, Kunselman LK, Slusarchyk DA. (1996). Enantioselective synthesis of α-phosphono sulfonate squalene synthase inhibitors: chiral recognition in the interactions of an α-phosphono sulfonate inhibitor with squalene synthase. *Journal of the American Chemical Society*, 118(46), pp. 11668–11669.

Index